IN SUDAN'S
CIVIL WAR TO
AUSTRALIA
LAW SCHOOL

Lual Jok Alaak

A Note from the Publisher

The publisher wishes to acknowledge and thank Dr Douglas H. Johnson for his invaluable help and support for Africa World Books and its mission of preserving and promoting African cultural and literary traditions and history. Dr Johnson and fellow historians have been instrumental in ensuring that African people remain connected to their past and their identity. Africa World Books is proud to carry on this mission.

© Lual, Jok Alaak, 2020

ISBN: 978-0-6450102-6-8

All rights reserved. No part of this publication may be reproduced, stored in a retrieval system, or transmitted, in any form, or by any means, electronic, mechanical, photocopying, recording or otherwise, without the prior permission of the publishers.
This book is sold subject to the conditions that it shall not, by way of trade or otherwise, be lent, re-sold, hired out or otherwise circulated without the publisher's prior consent in any form of binding or cover other than in which it is published and without a similar condition including the condition being imposed on the subsequent purchaser.
Design and typesetting: Africa World Books

About the Author

An officer to the Regional Defence Counter-Terrorism Centre for the East Africa Community prior to which he used to run workshops on terrorism and counter terrorism to Defence Attaches as part of their orientation on the threat of terrorism. He once conducted research on South Sudan Security reform with IGAD and has wider specialisations from law to counterterrorism as well as experience in South Sudan's Oil Industry. In his memoir, he explores his childhood experiences from the cattle camp of Pakou to rebel held town of Maridi and growing up during the South-North civil war of 1983, where he witnessed the brutality of the 21 year long struggle for South Sudan's independence. While a teenager in the rebel territories he trekked long distances through war ruined towns in the regions of Southern Sudan before he finally crossed to Kenya. Having witnessed the signing of Comprehensive Peace Agreement, Lual moved to Australia. He was once an observer at International Criminal Court in The Hague during the International Court Criminal Mooting in 2009 where his law school Bond University took on Yale University to earn them the world championship. In 2013, he returned to his country of origin where he works in various capacities as a specialist. Lual Jok Alaak is a lawyer and security analyst with a Bachelor of Law and Master's degree in Policing, Intelligence and Counterterrorism. He is currently a licensed advocate for all Courts in the Republic of South Sudan.

Acknowledgements

This book would not have been possible without the support of my wife Awel Kuol Mayen and children, family members of stepmother Nyanwut Malual Manyang, and the mother of boys, Apat Arok Deng and Arok Deng Achouth's family. Above all I wish to acknowledge my maternal uncle Abuoi Arok Deng who was like a father to me. I owe a lot to Ayual Garang de Alaak who took over my father's responsibilities as chief.

I am indebted to Deng Garang Bul who took me to Australia to interact with the best scientists and advanced technologies of the 21st century and some of the Ayual veterans of Koryom who negotiated me out of the Kalacha military training camp, so that I could go back to school. The list is inexhaustible as people like Nyanjok Chol Maker and Khot Ayiik Bul's contributions toward my success cannot be left out. I will continue to owe a lot to Professor Patrick Keyzer, Prof Laurence Boulle and Joseph Crowley for supporting me during my law school education. Without Barrie Hansen who kept the softcopy of the manuscript for ten years, this book would not have been possible. My thanks also to my village neighbour and friend Lual Reech Deng who has been encouraging me to finalise my book since 2008.

My gratitude and sincere appreciation go to the hardworking daughters of Kiir Monychol of Amour and Ajah who used to provide me with all the necessary reading materials with Ajah going through and making corrections in the manuscript. They exceeded all form of support by providing me with a car to speed up the completion of this book. My thanks goes to the management of Uhuru Cottage near Jebel Market and Majur Tulba Bol and his nephew Daniel Deng Makuer for providing me with unlimited power and a quiet space for writing.

I am always grateful to Akoy Machok Biar and Chol Yaak Akoi for we have shared every step of Juba life and to Eng Chirbek Marial who lent me his computer accessories to make this ten-year dream a reality.

CHAPTER 1

Aliet, my Birthplace

I WAS BORN IN THE UNDERPRIVILEGED VILLAGE OF ALIET in what is currently Jonglei State. Aliet village is poor in terms of modern facilities but is considered a land of fertility and natural habitats. As I grew up, I came to learn that my umbilical cord was cut with piece of broken elephant grass.

I was born inside a beautiful and spotless small mud made hut at night. There was no light, or a torch for some of us who were born at nights except bundle of dried grass ignited after two (*Akoy*) acacia tree branches are rubbed together.

Our time was not as lucky as when my little daughter Adhieu was born. She has a birth certificate. Her birthday is celebrated annually. I never had one. What is now in a record in the custody of my caring wife Awel is an age assessment certificate. It was made to facilitate my acceptance to formal schools and to formalise paperwork. It was calculated to allow me to venture into a wider world, a world which is not like Aliet village where documents are not required when one is traveling.

For instance, if one wants to rear livestock in Aliet surroundings there is no requirement by area chiefs to provide identification cards. Aliet could be easily roamed as a free ancestral land bequeathed to those people by God and beautified with acacia trees from its wetlands to beautiful horizons. There was nothing like daily harassment by power station companies calling my phone at six am from the Brisbane city centre urging immediate payment of electricity bills. The moon and the stars were our electricity: the natural gift of heavens. The skills of stimulating a cow's vulva after it had an abortion for the quick production of milk under pressure were essential skills for survival for young boys and girls at wut, the cattle camps. The technique of doing so, as I witnessed it at many cattle camps in the mid '80s would involve the pumping of air through the vulva.

In every homestead, people were taught about livestock management. We learned these practices not through classes but by doing it every day so that it became the profession of all the able persons in rural areas. That was how it was done. But suddenly we had to worry about things like birthdays and no one really knew when I was born.

My birthday on the record is guesswork. The computers as I came to learn are good for anything from fictional games, the wars and accommodation of entries to justify what was once natural village stuff. For us born in the village in the '70s and '80s, only a few had recorded birthdays. The lucky ones like Dr. Yaak Jurkuch Barach born in town had his time properly recorded in a certificate.

In our family none had a birth certificate except for my stepbrother Ayual Jr and maternal uncle Bior Abuoi Arok Jr. Both were born in Marid in the 1990s and nurses gave their mothers yellow birth certificates written out with pens.

Modernisation was not in many parts of the Sudan in those days. The marginalisation and poverty were extreme. Sometimes, violence if it did not spring from communal resource driven conflicts than it was either due to internal bloody conflicts or occupying foreign forces.

I used to attend spectacular birthday celebrations organised by friends in Brisbane. People like my cousin Akoy Machok Biar would struggle to remember when he had to organise his year-round birthday celebration.

The past of my people was miserable. Aliet village had no health facilities. That bitter past would later compound the complete absence of birth notifications in our villages of Ponborong and Patieng and Ajoung-Ayual. Against all these odds, families readily welcomed newborn children despite the countless number of sick children one might have had. With the birth of the child, the villagers would organise a gigantic birthday celebration. The celebration was an explanation of human reproduction and creation itself.

This means in all the celebrations I had attended I never recorded any slightest jealousy towards such festivities, however, I observed how rich people organised birthday celebrations in the emerging modern towns of Wau, Bentiu, Bor and Juba. Such events sometimes reflected the indifference of rich people to the cities' street beggars.

After all the trash and litter had been left on the ground, the end result perceived by bystanders is that a celebration by those who recently escaped poverty through various components of honest and dishonest dealings was extremely expensive.

In those days, before Sudan's Comprehensive Peace Agreement in 2005, our birthday celebrations were done by elders. The wine we used to drink was salivas. It was done by grabbing of one own's head before thick white salivas was poured on top. It was symbol of blessing. Then there would be a belief that child and the mother on whom the child depended for breast milk would go on their own long journey in peace.

A celebration with beer drinking reckless youngsters was not common. It was not common either in Aliet or Pakou, all the way to the Panget cattle camps. There was no hybrid wine. What was available had to be made out of millet and women made noises around it to attract consumers. It was wine purely for the council of elders. The children were only fed with milk, and fish and meat. Our bones would become so strong that boys and girls playing nearly crashed themselves. I would witness men who drank five litres of pure and fresh milk immediately after the milk left directly from the cow's teats.

In Australia like in other cities elsewhere, this was different. The young men and women to my surprise were drinking more of combination of alcohol and beer as if they were participating in Dimo Sudanese' lost boys' and girls' hot tea competitions organised by their caretakers.

Then by 11 pm, they would get too drunk to remember their way home or their parents' names. Everything valuable from their own bodies to anything in the wallet would become easily lost or unaccounted for. A friend of mine who got his Australian visa and had the privilege to be invited by his cousins to an end of year high school leavers' mega celebration could not believe his eyes when drunken youngsters began to distribute condoms to everyone like surgical gloves.

The hot tea competition was a new imported idea that found me in the Maridi Freedom Square in mid 1992. The Square was bordered with palm trees. The competition was being organised by Jesh el Aswod (the SPLA black soldiers) to bypass time before Commander Samuel Abujohn arrived to raise the New Sudan flag.

It was a competition of who drank the hottest tea first. It was easily for one to burn his or her lip.

Having seen the quicker assimilation of younger people to foreign cultures, Uncle and senior elder Bul Kuot Duot sitting under big tree had this advice for them. The boys were sitting on the top of three metre high rock and drinking Uganda waragi Kasese-Kasese (KK). He advised that people do not become winners with bottles of beer. In their days, he continued, happiness was based on number of cows one had.

In Aliet village those days, there was a life of unique excitement. Cattle left in the morning and disappeared from my sight upon reaching the thick forest of hundreds of thousands of different species of trees. It was an honour to my parents, and God that I lived to see this world and for them to wish me longevity, which was not only a blessing, a life changing experience. This blessing of longevity was in form of my aunt Nyandeng Alaak Ayual.

She was one of old generation of Pan-Ayual Juach with longevity. She lost all her teeth to old age in 1980 and lived to the minute of her last breath in 2014.

In the community, people like her used to compose songs and lullabies for new babies. The lullaby is how Dinka people welcome babies.

The first words in the lullaby start as soon as the midwife hands over a newborn baby to the mother and sound like: "Child, do not cry for this family has caring parents and blood related relatives. And its wider network of patrons and matrons would collectively care for you." As cited in the song, my parents did everything to prepare me to soldier through the thin and thick of life.

In Aliet, the simplicity of life and its communality was based on sharing identifiable activities and assets. For example, an empty bull horn would be shared around to use as a war trumpet or a wooden spoon use for removing out food from the calabash to a rounded horn embedded inside the grass thatched house to use as an urine pipe. The payment and sharing of bride price by family members and relatives would also include the sharing of traditional land, rituals and worship. Those social practices cemented the bond of Aliet community's togetherness. These were important in the prevention of family disintegration, which is now an emerging disease among our communities.

There was no social decadence as the bond was firm. Discipline was observed from the midwives dealing with the new baby to permissible visitors opting to view the newborn child. The discipline was that one's own mistake could lead to collective punishment. At the outskirts of Pakou, a leader could call us for flogging because one of the cattle camp boys had harassed a group of girls moving from Pakou to Patoot. The leader would act quickly that one could hear slashes hitting the buttocks at the speed of lightning.

As I am not like my little daughter born in Juba town; what else could I be? I am but a proper by-product of the village far away from tall buildings and medical doctors. This village has continued to see no tangible development as it remains as it stood since the day of genesis. Its inherent beauty is made up of the traditional practices and virgin landscapes.

Everything has remained untouchable, because the village has been at the mercy of traditional healers even before the 2005 class of V8 driving politicians began to promise the villagers with flowery words of development.

These promises later disappeared before their cups of tea they shared tea with elders dried up. Our village is like other villages in South Sudan. These villages have been at their natural states since the bygone six decades of Sudan's Independence. Its sewage storage is nearby acacia bushland. This sewage storage is available to anyone who wants to use it. It is free of charge. The sewage managers are birds, ants, the bacteria and hardworking dung beetles. This village has a simple setting which is sometimes exploited by politicians during the election to continue showering the villagers with once off gifts to induce them to attend their political rallies organised at Wangulei and Panyagor.

After they left, the longest saviours in the villages would still be gifted traditional doctors. These people have the skill of removing even the dead foetus from the cow's womb and mending broken animal and human bones. The people rely on their expertise to solve complicated problems.

CHAPTER 2

School under the trees in 1989

Having joined school at nine, which was later, interrupted by war, the teachers were asking for my birth certificate. I had no option but to choose 1980. This was after a request was put to my caring parents to ascertain the hour I was born ended up with too many events merging from one to the other. Everything including date and time seemed like it was being forgotten as my parents had been busy clearing land and restraining aggressive bulls, Mayom and Magar. My birth like the whole matter became past tense as soon as the placenta was disposed of.

The mess became ever more mystifying in the teacher's office. The teacher, Kuir Wieu, in SPLA/M rebel leader held area of Wangulei was in hurry in 1989. He needed my full name, date of birth (DOB) and next of kin. The reconciliation of events was made and problem solving seemed like it was working.

September was agreed upon. I was enrolled in class and I began to write my name by drawing it with pencil on the exercise books or printing it on the sand or wall with charcoals. The class was held under a big tree and was mostly attended by half naked boys. My first pair of shorts was open below my butt and was dark grey. The day my mom got a shirt and pair of trousers from Ethiopia was known in the village. I had to show my agemates my newly acquired pink shirt and blue trouser.

In the village I lived in, occasions were remembered by the appearances of notorious stars and supernatural dynamics. The clothes were totally alien to us.

Even an increase of frogs, termites, mosquitoes, water bucks, and tortoises had crucial significant meanings from invisible supernatural beings. If my mom wanted to send me for something, she would prefer to call me by my nickname, Madhier, the common name for a birth that happened during season of mosquitoes. I did not know the meaning until I was seven when she said I was born when a swarm of mosquitoes reigned and in a year referred in the village as the year of mud.

My affection with nature included observation signals such as delayed rains. The delay would immediately attract the intervention of rain makers considered by our people as stimulators of the rains (*Koc e Deng miit*). Then would come thundering rains accompanied by storms and strong winds which indicated the downfall of heavy rains.

As a result, terrified women would react by throwing grains into the rains while also protruding their heads out from inside the grassed houses because they feared being wet. I would just watch my mom and Adit Duot Guot doing it. It was a belief that doing so prevents the lightning from striking the hut or cowshed. But the force of nature is powerful and was going against any presumed practice when Abuoi Thon Kamich's cowshed was burnt by the lightning.

The next day, I saw about twenty elders with white rams invoking God (*Nhialic*) to adjudicate over their disagreement with the unknown evils that destroyed the shed. I could hear them saying in Dinka that justice should be as white as the ray of the bright Sun.

Villagers would read the signs to till their land. This was upon seeing the faster moving thick and dark clouds in April. The first rain to hit the village was that of tortoises (*Deng de Areeu*).

With its longest summer, the first rains were not enough for the crack prone black soil. The soil in April regenerates heat and hot vapours forcing hundreds of tortoises from their hiding burrow to the surface to gasp for fresh air. From here, I and Mayen Garang would pick the fat tortoises and put them directly into the fire. Then we would hear the shell cracking. We would sit and eat together the delicious tortoises' meat. The bitter part which could drain malaria out fast was around the anus of the tortoise. On the Guer Deng farm edge, other boys such as Deng Thuch Deng and Deng Mayen Ajak were also busy roasting both tiny and giant doves.

Other associated events hanging over the village used to assist people in foretelling and starting cultivation. These activities were done using ancient traditional hoes.

When the seasons for cultivation come, people like my lovely and beautiful Apat and stepmom Nyanwut, would not be exceptional in the village. There were many hardworking mothers I witnessed struggling with blacksmith tools.

Figure 1 My Mom on 1 November 2016 at Mumbai International Airport for her specialised treatment

As a child willing to study full time the hard life in the village, I would follow my parents as they went to the farm. Sometimes I would hear the giggling of my tiny bell tied around my ankle to alert everyone that I left our compound. This was done for my safety and to alert anyone so that when I approached small ponds, they would protect me from drowning.

I would go as far as Akoi Loy Adoor and Abuoi Thon Kamich's farmlands beating my bell with intention to learn what they were doing. Then I would come home to help my parents in carrying pumpkin like containers where seeds were kept. Inside the farm in the process of planting crops, I would watch Dad raising up his three-metre pole to hit wet and muddy ground to create a hole.

Then my brothers would put two to three of natural smooth coloured sorghum grains into those holes and bury them.

After a few weeks, I would call my mom to see tiny leaves grass like coming out from those holes. Then she would say those were sorghum plants. In June, Aliet is always in the middle of farming fever.

There was a lack of farming equipment and a common tool used for farming was Agot, a shrill ended hardwood. Agot as my maternal Uncle narrated to me in the day was quietly used throughout the centuries for cutting and burial. Its usage began to diminish in the 1960s when people began to interact with neighbours such as the Ciec people on the west bank of the Nile River. It was said the Ciec people had coal in their soil, which they used to melt iron to produce spears and other tools. The blacksmith would curve flat and rounded spears from it. Ever since the spears and sticks continue to be widely used for self-defence among the Dinka people.

The people without the tools would buy them from Ciec. The common form of trade was barter with cows, sheep and goats. It was time when people who have travelled as far as Mading Bor, Malakal and Wau following the first civil war came back home with a few metallic scraps.

Then the local blacksmith such as Bol Deng-akol Juach emerged as a skilled spear, cooking pan and fishhook maker. He used to make one to three hooks free of charge for me. Wooden boats then begun to cross the Nile from Ciec to Twiland enhancing the locals' river exploration.

The five kilograms bells tied around decorated bull necks were available for barter trade on Ciec land. Agot productivity was extremely low. For example, it could not clear larger acreages of land for farming. As a result, families could struggle as sorghums and finger millets were not enough to cater for them.

Embedded in the soil in the middle of many homesteads were hallowed woods for pounding millet. These were big smooth sticks women use to hit 30-40 centimetre holes of those woods (Dong) to crack the grains. Outside too was a sixty centimetres triangular shaped stone and a small tiny smooth round stone. I could find every woman and girl using these two stones to crush fine the fermented grains to make solid like porridge food for each household.

But still, the poverty was visible. Other means of supporting the villages such as fish nets were not in supply. I had the privilege of hearing Uncle Yith Deng Kuot telling stories to his kinsmen sitting under big acacia tree about how groups of people had to share one rounded spear during fishing excursions in those bygone days. The scarcity of resources was in itself a burden to the communities and the societies of that time.

But above all, my people were still remembering oral history which was dominating village unwritten records, riddles and folklore. In our evening sessions, riddles and folklores took most of our sleeping time.

Learning people's cultures and languages were so close to my heart. For instance, every day, I would leave home and walk barefooted to the outskirts of Aliet, Wangulei and Pakou. I would go to those places either to collect milk from the cattle camps or to sell milk at Wangulei or Pawel. On the paths, I would meet village members of the council of elders at four pm including my Dad.

Figure 2 Gak Garang Baak imitating his Lueth Deng's bull horn with boys at Athiel small cattle camp, 2009

Wangulei was where many attended to pressing matters deliberated under the trees. It was not rare seeing people quarrelling and leaving the court centre counting fingers from thumb to middle finger. I was not sure what that meant until one elder revealed to me that their mathematics of accountability of one's own livestock was not simple. He added that without papers to jot down each every single cow from unproductive, pregnant heifers, dogs, goats, sheep and hens, either borrowed or given to relative to pay off his bride price in turn in kind in the future, their ledger books were relying on memories. He was doing this as easy way of counting the number of lost animals. However, what was important were colours, marks and notches subscribed on the ears and noses of one own's cows.

Those marks were used in identifications of stolen domestic animals especially during the hearing of such disputes at traditional courts. Cases of stolen donkeys and hens were not commonly adjudicated among the Dinka people. Only cows had high esteem in both disputes.

Donkey and pig rearing was disregarded. This means one could attract laughter from the village. One would not take a month before becoming the focus of village ridiculing songs.

When elders decided to advise us while around cow dung fires, they would educate us on folktales. Others would say as simple as once upon time in their good days, there was good relationship between men and angels. That there was a ladder that humans used to access heavens. Then one day a woman who was pounded the grain hit a small blue tiny bird called Atoch in Dinka. Atoch became so upset that it cut the rope people used to ascend to heaven and back. As a result, we were cut off from heaven. What was left accessible for us was a beautiful brighter Sun which shines at the village horizons.

After I arrived at the cattle camp, I would think of how powerful the elders' expression of their youth, deeds or bygone days were before I retreated to my bed made of cow skins at ten pm. Paternal Uncle Adoor Majokreer was a respectable community elder. His wisdom was an inspiration to many people. His age was such that he could still remember everything his mudbricked made smoking pipe at Wangulei. Sometimes he would call us to help him search for it while running his hands around his ears.

We would say the important of those parts of human body was that it assisted him in keeping small rolled and pondered tiny wet tobacco as well. Then he would cause laughter when he said he became tired after his long day sitting in exhaustion at the Wangulei court centre.

From Monday to Saturday, chiefs from the area would adjudicate disputes over cattle thefts, adultery, child defilement, land, unauthorised drinking of other rightful fermented cow milks, aggrieved bodily harms, other injuries, rapes, stealing grains and divorces. The lists of complaints before them would be inexhaustible. That made their long day work tiresome. Memory loss was aged related and anyone could become a victim. Adoordit had tested his blessing, longevity, and his elder son told me that he was amongst those older than my Dad though they fell under the same age set of Mangok.

Some boys the wise men and women considered as disgraceful for their behaviour in the village such as those who ridiculed an elder staggering drunk due to the processed white grain wines.

The common wine in those days was made of sorghum. It could easily enter the blood stream to induce the limbic cortex of the brain to create happiness. Its taste is bittersweet.

At (wutbaai) cattle camps inside the villages' geographical limits, I witnessed cases where boys would sneak to Pawel and Wangulei to steal materials intended for making this wine.

But first the wine powder thief must meet some qualification from stealing small pumpkins from the farmers or beans and maize before the next promotion as ringleader. Whoever was caught engaging in these criminal activities would immediately trigger settlement disputes under big species of the acacia family.

The end result would later include flogging in public and a fine after the verdict is passed by designated community leaders. One day, Akoi Loy found us sitting around a fire lit for purpose of roasting catfish. He called us to attend. We stood up.

Then he started saying that anyone who does not possess altitudes of respect toward his or her parents, community social norms and culture would learn via the hard way. He added that we should pay attention to our riddles and folklores. He was carrying a hide made of bull which he used to reinforce the rules against the rule breakers. One of his arms was short. It had been broken by a Murle raiders' gunshot which nearly claimed his life.

He would also give us time to cite what he taught us from the days of Macot (tanks) to those of Ruon de Beny. He was saying that at that time, they were grown up warriors. He would continue to lecture how the local paramount chiefs were murdered by Arabs. What he was saying got later reinforced by my parents. My father was tethering cows inside the shed before he joined Akoi in conversation while we listened.

In their talks, I could hear Dad saying my elder brother Alaak saw his outside world in Ruon de Hon. William Deng de Nhial's assassination. That the month of May 1968 had been named after assassinated Southern Sudan politician, Deng Nhial.

Being the youngest in the family, the recollection I got from my parents was that I was born during the harvest. With its longest summer, the rains' onset begins in May and ends in late November annually.

Sorghum and millets as common crops, because of their friendly relationship with semi-arid weather, would be harvested before October. Then sorghum sweet sticks would later become wrangling resources for cows, goat, sheep and humans because it provides carbohydrates.

In those days, everything including the health of livestock and movement of fish in the rivers was left to God (Nhialic). And villagers would get busy taking the best of their harvests as offerings to the god of plenty, harvests, fertility and rains: the god of Deng-abuk and Garang.

The importance of supernatural spirits among the villagers was demonstrated in *Atemyath* (African adder of shrine). Atem was a small god of Awulian people and had its shrine in 'The *Luangateem*' east of our homestead. The east is considered by the villagers as the south (*Cien*). That is to say they believe that the Sun rises in the morning from the south. Then in the evening it heads northward before pulling down at five pm at the horizon of what constitutes the Pakou edges.

Atem is an African adder species considered as sacred. Its connections with people was evident when villagers fed stray Atem with melted cow butter. This ritual was done to make a peaceful send off had Atem moved itself into cow byre or some house by mistake.

All these butter and cheese gifts were going to waste. In 1986, I vividly remember, my second elder brother Ayual was nibbled by an adder. He was in serious pain and his leg was swollen from the ankle to below the knee. The family was in deep shock. There was no medicine to treat the venom. To comply with certain beliefs, I was called out from the hut (*Hot*). My father gave me instructions on what to do.

He brought in the fermented bark of acacia tree peeled from *Thau*. This Thau made foam like soap. I was given strict instructions to wash the swollen foot with liquid foam. I was told to run the foam from the knee down to the toes. The foam had herbal properties. It was capable of removing the venom. After three weeks, Ayual became better and left for Wut. It was presumed that some species of acacia contain psychoactive alkaloids, potassium and fluoroacetate and can be used to poison rodents. My Dad was capitalising on the good use of the acacia tree's properties.

Chapter 3

The traditional snake beliefs and offerings

With the shrine located deeply inside Awulian heartland, the offerings to Atem were brought by most of Twi people from near and far. After the villagers left the altar, no one would say exactly whether the men in charge of the altars became the final beneficiaries of sacrosanct items or not. Our villages were that the rumours spread that the masters of shrines were alleged to have always been rich as materials left there were not going directly to the gods. No one knew whether the allegations were true. But like in agency businesses, they were people in the middle.

With offerings to Atem either he asked it directly from anyone from Twi or volunteers. The biggest offering was to ask people to provide a bull. Malual Longar who passed by the shrine looking for his stray cow was informed by the masters of magic spears that he got favoured by Atem. And that it was the Ayual community's turn to provide the offering. With instructions from the divine god, he quickly took the message to his age set.

Without hesitation, he left what he was up to and took the news to the Makol de Ayual age set. Everyone argued with him bitterly that he was responsible. That he should not have gone closer to the sacred shrine. After a heated debate, he had to do it as they insisted. Then Ayual people made offerings according to my parental uncle Kuany Alaak Ayual.

After few days, paternal uncle Kuany continued, the villagers saw a giant elephant cross to Ayual land from the direction of Kongor's land.

To confront the elephant, the community organised strong men. They were mobilised to fight the aggressive lonely elephant. Among them were the sons of Malual. It was a custom that whoever speared and killed the elephant first took the tusk. By luck, the winners were Malual's sons who later sold it for a reasonable number of cows. They became rich. It was assumed by everyone in the village that Atem had compensated Malual for his generosity. In Africa and as in the Biblical and Islamic teachings as well, people who give receive in equivalence.

At the altars, anything of value was left without questioning the beneficiaries. It was like looking for a curse for one to ask what they were giving out voluntarily.

Then one would just burst in a song praising anything of his or her favourite available on earth, or any other such imagination of its kind in heavens, while making a quicker retreat from those guarding the sacred places. The issue of accountability on how those valuables disappeared after successful delivery was left to the invisible spirit to decide.

At such a time, anyone whose birthday coincides with plenty of crops has come at time when stomachs were expected to be full from family members to the workers at altars. Stomachs as Dinka say are interchangeable with aliek (food pipe) and have their own gods: food. They worship food. Death is imminent if one goes for several days without eating. Legend singer Panchol Deng Ajang de Luk has connected happiness with veins and arteries as based on someone getting a constant sufficient supply of blood from properly nourished sources.

One cannot let go starving guests without pushing away his fortunes. It was for this reason many people in the villages talked of a man called Makuach Ajak. They would say Makuach erected his houses where enormous foot alleyways converged to enable him to feed exhausted and hungry strangers free of charge.

CHAPTER 4

The Youngest Son

As the last child born to Jok Alaak Ayual Juach and Apat Arok Deng Achuoth I had the privilege of spending my good times with my parents. I had to enjoy other privileges coming with them, and the surrounding environments. For instance, I was able to know when the time for planting and harvest would be. Then Dad had his idea. He had in his heart that the time to honour the Arok Deng family was at his doorstep. I was allowed to inherit a name derived from the Adhiok section of Pakoy.

The community of Adhiok in Man-Deng brought my name to Pan-Ayual Juach. The team was led by my maternal uncle and senior elder Mach Dit Akech.

In accordance with our custom, I was kept inside a grass thatched house for months away from the view of the public. Thing were not like nowadays where new babies easily find their innocent faces on the Internet for a quicker unlimited view by the world. In Dinka, girls remain sources of wealth. For instance, they provide cattle and other form of bride prices or other unrestrained supplies of valuables from the groom sides. Then there was this idea of these cattle camp boys being considered as warriors, protectors, and the inheritors. In the event parents had a blurry idea on how assets especially cows would be divided in the future, the boys' shares would be made clearer. At maturity, the first victims to be excluded from family inheritance would be the girls. The younger male in the family as well would be instructed to share the remainder with their mothers.

Without a computer scan as my parents could not receive information before my birth, everyone was waiting for the delivery outside, with some inside like Mama Adit Duot. Then it was another boy in the family of three boys. I came out when every nearby villager was glancing through small drilled windows. Their aim was to see or hear what the outcome of the inspected genitals had determined. Then I began to breathe the fresh Aliet air.

The villagers who had come to witness the birth the next morning left and it might have been possible for the number of boys to rise to four. That was a turning point.

Years before my grandparents had one son. It was my Dad among the four sisters of Akuany, Nyandeng, Amou and Achol. This time, there was no sister. In Dinka society, those days, boys were valued more than girls. The pressure families with daughters without a boy faced would force them to invoke everything from deities and to the father of all gods to give them a son.

The only sister Akuany had died earlier. The whispering from the bypassers might have been that the nhialic of stomach should provide any luck. It is because a house without a sister was a boring house. In such house, every boy had to soldier to survive.

Then there were other challenges of women struggling to collect firewood without any girls providing extra assistance. That woman was my Apat Arok. The Dad age-mates and stepbrothers of Atem Alaak Atem (Adeen-wut), Adoor Juach Loy (Majok-heer), Garang-marol Juach Loy, Kuany Alaak, Garang Alaak, Juach Alaak Kuot (Mapiordit), Lueth Deng Baak (Gutawet) were revealed to have paid family a visit on that date. Coming with their blessing too were my maternal uncles having walked long distance from Adhiok-land.

I vividly remembered my maternal granddad Arok Deng spending time with us too at our cowshed in the mid '80s. All those activities had not happened without anxiety.

One was a genuine concern of a house without sisters. The issue of bride prices was removed out of equation as we reached the puberty age. In those days where the best standard of lives in the village were measured with cows and fishing, our premeditated future was anticipated continuous fishing sprees. Another concern was that one or two boys should temporarily look for odd jobs in the city of Khartoum. That one would work there as a porter or brick layer.

If one does not look for cows, getting married was out as only a few good-hearted people with trust would give their daughters to a man who is poor. The payment of bride prices was by cows with a number not less than thirty. Paying anything less was considered credited against someone empty's cowshed. Other chattels such as goats, axes and spears had lower acceptance rates. Not even dried, or fish caught alive would come closer to the bride prices.

There was tension that people had accepted the challenges of having to mobilise enough cattle to be used as bride prices. The grown-up men were under the pressure of being forever bachelors.

To refuse to get married in the village was a serious crime of one own failure. The challenges one would have to stomach would be failure to rescue the village from extinction.

People were expected to produce in abundance. Maintaining the family tree by getting oneself a wife and producing children was among the top lists of community priorities. Under the trees, I would watch some marriage negotiations either being settled or deferred for next time's clearance.

Complete failure or default after several oral notices passed to the other side by the elders of the girl attracted confiscation of the children and the possible changing of their names to that of one's father in law. So, paying bride price was must. Failure was associated with disputes awaiting ahead in one's lifetime. Those who worked in big cities like Juba, Malakal and Khartoum were said to be sending enough money home, not to cater for village remaining parents but also to contribute toward their bride prices. In return, they would buy enough cows for their needs.

Toch was also good place to stock enough dried fish for market. With barter trade being pushed away slowly from the village by monetary trade, everyone in the Pawel market was going for money to money exchange. There were auctions held where all forms of livestock were sold.

For a family without daughters to accumulate many cows they must unearth a livelihood out of hardship. We were destined for that life. However, Alaak at the age of ten was beginning to show excellent skills in hunting.

He would criss-cross deep swampy areas and his spears missed no fish, tortoises and monitor lizards. He rarely came home empty handed. He would come with enough stocks of fresh fish to feed us for weeks.

He had already fully accepted the biblical saying of human must eat from own sweats. There were times he would dirty having carried home with him those high in protein mud fish.

He is left-handed like his grandfather but also with hunting luck as known by many in the villages according to testimony from the gifted traditional singer and skilled hunter Akoy Deng Chol and his hunting mate.

He was so good at the art of making food from the bush sources that he would leave as early as six am and come back at six pm. His lunch was based on how best he used his spear. In his absence, I and Mayen would make sure we cleared cow dung and looked after calves.

Figure 3 2019 Pan Jok's family reunion with Ayua Sr and Jr, Achol Madiing, Awel Kuol, Jok Jr, Reech Kuol and with family kids
Courtesy of Ayual Sr compound

Other than boys' burdens, there were also other burdens in families made up of exclusively girls. These burdens were not limited to accumulating herds of cattle without heirs. These girls would have to do the boys' harder jobs, such as looking after cattle and search for stray ones in distant cattle camps and being chased by hyenas. These girls would befriend huge and fat boys to protect them from thick-headed bullies. These favours would not come without sexual abuse. When these girls are married off in the later time, the family assets would be exposed to crowds of standby predators.

God works His own miracles. We grew up as healthy boys. My brothers Alaak, Ayual and Mayen with significant names proved themselves important among their peers. Ayual was good wrestler and Mayen was best in stick fighting which scared those who enjoyed robbing little boys of their fish, wild fruits or turtles.

I would watch them in excellent stick fights and one to one head-on opponent wrestling.

As matter of cordial relations, and good gesture to our ancestors, our names were spread across. The first name was for my parental linage. My name was drawn to maternal uncles in the legend of Lual Akoy de Mier whose thousand descendants continue to spread wider.

Ayual and Mayen who were in the middle had names. For instance, Ayual is named after grandfather while Mayen had connection to a bull made as crucifixion to small god *Deng-Pakeny*.

My name was considered as the best gift brought by elders from Adhiok. Food was brought and other important items. The joy of receiving such gifts might have helped relieve my mother from the ordeal of labour without painkillers. Like many women in rural area, she had a normal delivery. No doctors, no medicines. Pongborong had everything natural which makes it unique from the rest of the world.

In African societies, naming after great grandparents is instrumental. My father derived his from a sacrifice made so a family could have a son.

All forms of supernatural interventions were invoked. Then he became *Jok* for spirit. It was believed any prayer of positive outcome had consideration as the works of the spirit of good fortune.

The rest of the names in the family came from Ajoung-Adiang. Guer who begot Juach, and Alaak and hence section, Pan-Alaak. Juach was married to Amon Diing. Then he decided to settle at present day Ayual land to live permanently with his in-laws.

Amon begot her son Alaak which later constituted the present day Pan-Alaak. As a community of people, Pan-Alaak continues to grow. Amon was the elder daughter of Diing. She was with her stepsister Anok at those immemorial times. On the advice of her father Diing, Alaak was assimilated to Yak circles to form Wut-Pan Yak with thousands of descendants. They have ever since constituted part of the modern day Ayual community. Their pride and legend have been cited throughout the generations. Brave warrior Baak-Tod and their local war strategist Atem Adoor were widely respected and known by their peers in the 1900s. Community songs including war songs were composed in their names.

Alaak later got married to Achol. Pan Alaak born to Achol is made up

Figure 4 cattle camp boy with his bull near dung fire

of three sons namely: Baak, Adoor and Juach. Their sister Angiec later got married to Kongor Anyang

Ever since, they have fused with the people of Akoy Anok (Angui) and their maternal uncle Yak Diing to jointly constitute three major sections (*Wut*) of the Ayual community.

They eat together. Also, they fight as one set to defend themselves against attacks by adversaries. These people share a lot in common though intermarriage. This intermarriage was made possible to cement the bond.

To ensure they multiple and expand, Yak called his descendants including those from Alaak and Angui.

As a covenant, Yak Diing slaughtered bulls and crashed bones. The crashing of bones has ever cemented the intermarriage among Yak, Amou and Anok siblings. Without that bond, intermarriage between Amou, Anok and Yak descendants would have been prohibited because of incest had it not been before the covenant. The arrangement was done during the early onset of Ayual people and the year is not known. The Ayual community consists of hundreds of thousands of people. Its early settlers according to its oral history were Ajak Kur Ayuel and Ajak Anguac. The former was from Roordior and the latter was from Achath. The legends contributed massively in the settlement of majority of people not only in Ayual's land but as far as adjacent vicinities. This earned them/Ayual the name of uncle of Wuts. The two Ajaks' historical sites (*dhien*) remained inside the heart of Pakou.

The current sections, which constituted Ayual people were organised and allocated lands by the duo. This was during the early resettlement and settlement of people and this could have been around the 13th century following the disintegration of the Nubians' Kingdom of Alodia which stimulated the exodus of the Jieng people (Dinka). The Dinka people according to history were said to have migrated from Gezira of modern Sudan southward. Many factors such as adverse environments, conflicts and raids pushed and spread them across the modern-day Dinka-land. Currently, they are the predominant residents of Greater Upper Nile and Bahr Ghazal.

Here, the two Ajaks assimilated many strangers into them and their influences increased. Having convinced more seasonal people to stay and settle permanently with them, Awulian's people cited their migration from Bahr el Ghazal. They crossed the River Nile to present day and offered a ram to Nyuak. From the hair of the ram, the name was derived and Nyuak which was origin of name of modern day Ayual was agreed to be shared between Dachuek, Awulian and Ayual. The Awulian had with them their god, the snake Atem.

When human beings were regrouping to form social units, Ayual was also busy forming itself from Wut-Roordior, Wut-Pan-Yak and Wut-Pawiir.

As three pillars, these three branches have remained as one entity for hundreds of years. Their largest cattle camp center in equivalence of a modern state city is Pakou.

Inside the center of Pakou are Dhien or piles of tiny ashes from cow dungs from each unit. They are about half kilometre squared in size. This is where all Ayual in those days made sacrifices to their god Wieu. Wieu is a grass snake. Spiritual rituals to Wieu were sometime done annually or on it request at will.

Pakou even today is where people settled their cows year-round before Ayual leaders decide to move cattle and people to the wetland Toch. Moving to Toch is done in summer when the scarcity of green pasture is observed.

But no move shall be made without rituals. For instance, the ritual would be performed at Pakou. After everyone has agreed on the next move, the first section to move as cattle leave Pakou in the morning would be Pan-Yak (Pan-Wach).

This exercise is done early in the morning. Upon reaching the outskirts of vase wetland (Toch) which comprises of Wut-e-liet, Buuradut, Ariai, Acinyol, Buweng, Kathiac, Parioth, Nyal, Akoy-de-anyial, Dhiop and Bioth-agany to mention a few, other rituals calling for safety of animals and humans, before crossing the deep streams would be performed. Everyone would look up at the Ayual section of Pawiir for assistance. Here, they would invoke powers of their god, Akoi, to bless the movement of people and livestock.

Akoi was considered as god of the river. Its functions were not limited to determining fortune and anything related to the rivers but also protection from waterborne diseases. Toch which comprises of water and thick Sudd vegetation is home to the hippopotamus, crocodiles, snakes and strong currents. According to the beliefs, the rituals were to make them friendly or peaceful. The men of rituals would ensure their teeth are powerless to the level that they could not bite people and cattle. Like most of pastoralist communities in Africa, the community of Ayual has generational age sets going back to time of its composition. This generational age set included the Ayau, Makerlien, and harkeen, Adhieweng, Makol, Chol, Mangok and Madiing.

Figure 5 Ayual Generations as narrated by senior elder Uncle Ayiei Bul Ayiei Pajieth
Courtesy of Ayual Jok Jr

My brothers including the stepbrothers have remained as part of Madiing. Decades have passed since the last Ayual age-set Madiing was initiated.

Even my sons Ayual and Jok if an age set is not initiated would also join Madiing.

For worse, those who are living in town do not even know anything about it and this cultural practice is on the road to disappearance. The Awulian community tried it recently at their Pawuoi, which was well attended by personalities of different views with the rest addressing purely Dinka conversants in foreign languages. One elder having seen the future of Dinka culture was shocked by the level of cultural cross breeding. He concluded that such intermarriage of western, Arabic, East African cultures with theirs was his major concern. He took his stick and left. But he was monologuing that people nowadays in their beer places condemn their own cultural practices while praising everything white people had left in the books.

In most of the communities, the proper organisation and naming of our age-set was interrupted by war. Before Chol, there was Adhieweng. This age-set fought in enormous communal violence to defend the land even with neighbours. Then people would be lost on both sides and community and in one incident with Kongor, Ayual elders would accept that they were displaced to Ajoung-land and back to Dachuek-land. Then they made an initial decision to migrate to Mundari land. Leaving the lands and Toch was not acceptable and amicable solutions to the conflict were reached by the like of the then community leader Kuer Bul Duot and these two brotherly communities have lived peacefully and in harmony for a hundred years. They shared everything in common from marriages, sharing of grazing zones and adjudicated border related issues together.

In mid '60s, people were hit by severe floodwaters which lasted for years. Thousands of people were displaced and cattle died. It was natural disaster of significant history, which continued to be cited in many oral statements. As a result, the majority of our people in Aliet went to Bor. The rest went to Aborom and finally, Nuerland. Aliet was submerged in water and impenetrable vegetation of all kinds of water borne grass and weeds sprouted up.

My Dad having lost his father at young age had to move their cattle to higher ground at Borland with his sisters and grandma Akuany. Then my mom was also with her maternal grand uncles of Aguer Bul Aguer family until they got married around 1965. Many villages including my mother's birthplace Piompur were desolated land. Just before the beginning of the '70s, people began to go back to their original villages.

Aliet because of flooding was not conducive for dwelling and livestock.

Then my parents decided to settle at Pongborong.

They stayed there for more than two decades though it was painful for one to abandon ancestral land. This was because our home at Aliet was where all the great grandparents were buried, and our parents could not afford to abandon it completely. So, Dad was not happy to have his family stayed at Pongborong. Pongborong was also far from Pakou which was acting as Wut Yath meaning a centre for communal rituals.

Finally, our family with that of Garang-marol Juach was moved to Aliet in 1977. Years later after I grew up, I began to pass via our former village while heading to Penyde Leek, Piomahol and Patoor. This was the time when I decided to pay a visit to my aunty, Nyandeng, who was married to Reech Akol Reech. As a result, I grew up with the privilege of knowing the strong connection our people have to ancestral land. Then my parents used to lecture me on the two sides of familial lineages. In fact, our people respected blood relationship and one must know both sides.

This was important in the exchange of important issues in life. For instance, meats for marriage were divided into pieces. Then distributed amongst one's own descendants.

Bride prices were also shared among enormous numbers of relatives from the first to the last cousins. So knowing one's relatives helps in many forms.

Figure 6 Alaak Deng Manyang cows during marriage negotiation May 2019 and looking on are Aliet elders and women

For instance, it protects one from curse in cases of corrupted dealing in distributions of what community considers as items meant as theirs.

In this practice, certain meats and bones were dedicated to cement one-to one bonds. Marrying among those perceived as related was also discouraged. The reason was barrenness if one breaks such bond even if such relationships goes beyond the sixth cousins. Child naming was also important.

Great names of well-known personalities continue to dominate the Dinka cultural and heritage. For my brothers, the celebration was for the first batch of three boys. However, the family was still smaller. In our family, there were few children before I joined the crew. This news was said to have included a celebration in anticipation of family expansion. Sizes of families were so crucial those days. People would fight and the lack of manpower to do so meant one's own farm could be easily encroached upon.

In 1987, I saw expanding of extended families in the village as men with wealth embraced polygamy practices. In our family, we had death too. The only one family's daughter Akuany was no longer part of our oneness. She had unfortunately passed on at infancy according to my parents' account.

Any child after the deceased was regarded as compensation for natural injustice and would take either Chol (boy) or Achol (girl) name. The loss to us was difficult to replace. The late Akuany derived her name from the Pawiir section of Pan-Yaak. I vividly remember these people. Dad used to send me to pick cow meat reserved for Akuany siblings. Taking meat not meant for you during the dispatching causes dispute. I would grab exactly what I was instructed to and then ran back home. I would make sure hawks were not following me.

Hawks from the subfamily of accipitrine would fly lower and snatch what one is holding. To scare them, I would light a three metre cane and hold it higher while it was burning on top. That was enough to keep them away.

They have a habit of hunting by flying down suddenly from obscured perch. They prey on hen, fish and rats. Their powerful force could break someone's fingers.

Once busy under the tree peeling filaments around the giant bone was Madany.

Hawks were following him unknowingly. It was summertime and our stomachs were empty. Hawks were also looking for what to steal.

In a blink of the eye, Madany was robbed of his meat by the large hawk. This action of the predator did not go well with elders sitting under the tree eating meat from the bull side slaughtered to seal the marriage of my uncle's daughter Lueth Garang Alaak.

The predator was condemned and in few seconds, it crashed. We went and took the meat. In the distribution of meat, nowhere was it the hawks right to share from such marriage. In payment of bride prices for instance, it is only the blood relatives who offered the cows. Not distinguishing what did not belong to the birds, and what belongs to rightful owners had cost the life of the snatching hawk.

As a family, we had to accept the loss of our sister Akuany and Panther named for ancestral olden home (Juach) than persist in the grief. The family was in period of bad mood however, the decision was left to the creator, God to bless the family again.

That meant, by recognising the family's heritage and shared names with uncles and in laws, Dad was doing equity to invisible ancestors and sought blessings by doing so. Ancestors in traditional African societies were highly regarded as living spirits who manifest themselves as the living dead in various forms.

There were times when we would be scared by villagers' assumptions that night orgies represented skeletons of the dead. This assumption would be continued that they come out at nights and wrestle with the living.

They would say that anyone defeated by this ghost disappeared. Scared of this, I would make sure I was at home by five pm. While looking after my calves, sheep and goats, sick cows, I would make sure I read the signs of the sun going down. The reed made door of our cow-byre was facing east. So in the morning, I would watch the bright sun rising and settling west of it.

I would keep in mind the challenges of coming home late with cows when hyenas made their presence known with sounds. Also, to worry about were cobras and other snakes.

Looking after herds of cattle was a job done without shoes. In the rainy season, the whole season would end while one was being rained on.

In the evening Dad as the chief of cattle camp known as Benywut would come to where he sat about four meters from the hut thatched kitchen and advised that he had done his part by raising us and had made his name as wrestler and gentleman among his peers. In March 1987, the possibility of family being set apart by civil war was imminent.

In Sudan's Civil War

The Sudan was four years into civil war by then. The war was pitting Christians and animists south against Arab dominance.

Those able men but older than my brother Alaak had left previously for the Bongo training camp in early 1984. By September, the men who left behind decorated bulls and their fathers' riches were armed combatants.

I could vaguely remember the like of Mapiou Nyuon Abui, Magot Akoy Bior, Adol Deng Baak, Chol Deng-Akol Juach, just to mention a few the day they abandoned every precious thing in the cattle camps. They left the best life to form the best fighters Sudan had ever provided: the Koryom battalion.

By 1985, after nearly two years liberating areas in the South from the Sudanese Army Forces, bad news of those who perished in the battlefield began to reach deeply into villages. As young as five, we would run to any homestead when we heard people wailing.

Then it was nothing other than painful and shocking news. The sorrow of families without opportunity to bury their love ones was terrifying.

One early morning of November 1985, I was sent by my parents to fetch water from Wangulei, the only tap installed by the then Southern Regional government of H.E Justice Abel Alier Kur.

Just as I left the edge of what once was the Dau Juach (Dau-gutjok) Loy farm, I saw hundreds of men in military uniforms carrying big guns the size of sticks used in the villages to pounder sorghum and millet grains. Some of those guns had to be carried by two, or more men.

I was terrified and had to retreat home quickly. I told my parents what I saw and Dad said that what I saw were fighting (kake tong) weapons and Anya-nya. In most of the resources motivated communal violence, I had witnessed, spears, batons and shields were commonly used. Guns though few were made 80% of wood and rarely got used. Spears and big sticks were occupying bigger portion of age to another age set stick fights in the villages or in temporary settlement at cattle camps.

This was nothing like that I had ever seen. After two days, Chol Deng Juach whose house was adjacent to our place made a cross where our farm joined that of Abuoi Thon Kamich. How he dressed made it difficult for us to recognize him. His mother who was almost blinded recognized his voice.

They could not control their joys and the gun fire ensued. First we were terrified and the young ones from among us such Juach Abuoi, and Thon and children of Akech Adier fled to nearby shrubs near their farm startled. Having seen me standing watching as the event unfolded, they had confidence and made it back. We jointly started to collect aggressive case tapers as they dropped from the AK 47. Those aggressive case tapers were empty and comprised of wide bore diameter, short cartridge overall, wide extractor groove and thick cartridge to enhance extraction functions in all conditions and allow higher extraction velocity.

Some of these soldiers were notoriously wild, as they were taught that even if one's father refuses to obey the liberation objective, shooting is permitted. They had a song to justify such actions. People were scared too as fighters would use force to bring everyone to submission. As a result, some would take goats and bulls at will to make dinners and lunches from.

Villagers rarely talked even if one recognised their own bull under knife. The soldiers were tough. As a result, hundreds of goats, sheep, and bulls disappeared mysteriously. This news of lack of discipline immediately reached the High Command and Commander Kuol Manyang intervened.

It was not all the soldiers who were not disciplined. There were others like the highly disciplined soldiers like Chol who would seek permission from the cattle camp head chiefs to provide them with food and little milk.

This would be where people like my Dad would come in to feed the soldiers with what was reserved for starving children handed over to them in a dignified fashion and with respect.

Any child who tried to protest why his or her designated cow's milk is taken would be told to keep quiet or Anyanya would be called to do the beating. The villagers in such event would do everything for the survival of the rebel movement. Dad would go the extra mile to give out his bulls and a heifer as a chief.

The first heifer was white and last bull was black to signify our land. In the ancient time, anything for offerings must have significance before it is made.

Here, the black represents people and Sudan in general. He did not tell everyone why he did that until a year later when he explained to his stepbrother Kuany Alaak.

Also, there was the issue of anyone defying the orders of Anyanya had to be properly punished as villagers easily confused them with Anyanya One a rebel movement formed in 1955 one year before Sudan independence. The one in the village was an SPLA. Its leaders were Dr. John Garang de Mabior, Kerbino Kuanyin Bol, William Nyuon Bany, Salva Kiir Mayardit and finally Arok Thon Arok. At the onset of the movement, they had over 12,000 fighters, which later increased to over 25,000 to include Koryom and Mourmour battalions. The rest were former soldiers of 104 and 105 absorbed into the Sudan Army during the Addis Abbas Agreement of February 1972.

The way SPLA were treating own comrades, who defied their senior officer orders were brutal. While seeing this, we would say if they could act tough on their own comrades in struggle, who else would guarantee that non-combatants would be spared. Defiant chiefs began to receive flogging.

There was worry but the soldiers later provided all the needed protection from the enemy. War planes which used to bomb our livestock and farms when we took them to drink water at any water points became scared.

These weapons I became familiar with years later as surface to air missile-SAM7 would be fired against passing hostile planes. Other weapons were the RPG or Rocket Propeller Gun and Degtyaryov machine gun.

The Koryom song was that this machine gun would shoot cowards from the buttocks if one retreats from enemy line. That meant it could go very far, different from AK47 and one must not remain behind during the close range assault if to escape from its pitiless wounds.

Mourmour and Koryom battalions were the deadliest against the enemy. They would attack once and for all without any form of reservations. Like the Sun Tzu theory of war, their victims would be left properly destroyed and annihilated completely.

Then the news of the people's movement's achievements was celebrated in the villages.

Everyone would dance from the leader John Garang to cattle keepers in the forest upon hearing the fall of big towns such as, Kurmuk, Jekou, Nasir, Rumbek, Kapoeta, Torit to mention a few. The prisoners of war of both Arab origins and Africans would be paraded and cared for. Those who surrendered themselves from the Sudan Army were enormous.

Herdsmen were temporarily relieved from Arab tax collectors. The machine guns operations were done by muscles and brave men who were accurate and well trained.

The group of comrades Chol were under Commander Arok Thon Arok. They were on retreat orders as some of them had died at Tingling desert because of thirst and fighting or they were the survivors of many assaults against over ten thousand Sudan Army soldiers' defeated by them between Jemeza and Bor. This fight was considered in Arabic as the battle of ashara alef.

Few weeks before the Christmas celebration in December 1985, Chol and his colleagues left the village. Some since they left us perished in the line of liberation duties. But in Pawel and Panyagor, I would find few soldiers loitering in the town but waiting for orders. Dr. John Garang and his command had assigned commanders into battle-axes, or war-zones. Sudan Army outposts such as Poktap and Bor were shaking.

These towns were besieged and in constant fear having suffered from SPLA bombardment and gunfire. The sounds of guns were shocking and thundering. Terrified by the ruthlessness of the SPLA, the enemy which was deployed at Poktap fled.

As a result, Poktap was liberated and I could see smoke covering the ground and sky. Giant machineries used for the digging of the Jonglei Canal were planted with landmines and blown up. Again, there was dark smoke different from the burning savannah grass.

Madiing Bor as it was called had to experience a second attack different from the 16 May 1983 dawn attacks organised by Maj. Kerubino Kuanyin Bol and his soldiers. This town was attacked again by properly organized SPLA fighters and liberated in 1989. By the time Bor was attacked in 1983, it was fighters of southern origin in the Sudan Army force. The second assault operated under the banner of the SPLA/M.

Here, it was people at war. Dr. John Garang De Mabior as a rebel leader had mobilisers. People like Commander Chagai Atem Biar and Adoor Arok Akol had mobilised Koryom and shortage of manpower was impending. Other notable commanders in Bahr el Ghazal and Equatoria regions were doing the mobilisation. Recruits began to arrive in our villages. They were so friendly and without food, they would ask for help from the villagers. Some would come to our houses and got little help. The villagers rarely had anything. Some would help us cleaning cow mess and then the next day, I would see them no more. The sick one would remain there in the care of the villagers until they are healed and then joined the crew. Having crossed vast land from the Bahr el Ghazal region all the way to Bongo, or Bilpam in Ethiopia was the longest journey meant for men who had decided to give their lives for freedom.

It was not an easy journey on foot with ambushes from pro-Arab militants while crossing the land northwest of modern Jonglei State. Sometimes it was difficult for me to understand their accent despite them being the same Dinka people. The Dinka of South Sudan is the largest tribe of millions of people. The Dinka who referred to themselves as Jieng have about fifty six clans namely the Rek of Wau, (Dinka Marial

Baai), Palieupiny Malual, Pajook Malual (Malual Gieernyaang/ Malual Buoth Anyaar), Paliet Malual, Abiem Malual, Twic Mayaardit, Kuac Ayok, Awan Chan, Awan Mou, Wan Parek, Aguok Kuei, Apuk Tonyrok, Apuk Padooc, Apuk Jurwiir, Kongor, Abiem Mayaar, Lou Ariik, Lou Paheer, Luac (Luanyjang), Luac (Luanykoth), Akook, Thiik, Jalwau, Nyang Akoc, Abuook, Atok, Noi, Leer, Muok, Yaar Ayiei Cikom, Thony Amou Marol, Gok, Kuei (Agar), Rup (Agaar), Pakam (Agaar), Parial (Agaar), Yak (Agaar), Atuot, Ciec, Aliap, Bor, Twi (Twi of Jonglei) , Nyarweng, Hol, Luac Akok(Luac of Khorfulus), Rut, Thoi, Ruweng Paweny, Ngok Lual Yak (Ngok of Malakal), Dongjol, Nyiel, Ageer, Abialaang, Ruweng Paanaruu, Ruweng Aloor and Ngok Jok (Ngok of Abyei).

These clans like other tribes contributed massively toward the liberation of South Sudan. In their own uniqueness, the language had varying enunciations.

To achieve the objective of second recruitment, chiefs were persuaded to join in support of the ideology of liberation. Political emissaries were posted in town nears the villages. Intelligence officers were drawn even closer to the villages to get the names of those against liberation from Arabs. Then people started to call themselves comrades though some people would joke that anyone considered a comrade is someone who takes other people goats without permission.

Chiefs like Deng Ajang Ajak (Dengayang) had to identify families with boys to be conscripted to the rebel ranks and files. This process was (Buluk ke dhiak) - chief with three volunteers. We had three chiefs with Deng Ajang Ajak as a head.

On 3 March 1987, my elder brother Alaak was approached through my Dad. He was looking after our cows at cattle camp Akuoc when he was informed to come back to Wangulei. He was not alone in the list, because he was with Atem Deng Ajang, Deng Guer Deng, Dhieu Mangok Alaak, Ajak Deng Lueth and Chol Bior. At Wangulei, they were joined by hundreds of men.

At the top of our family list was Alaak. At about three pm, Dad had called Alaak home and informed us of the decision. The meeting was done inside the cow byre. Then I listened carefully as it was not pleasant to be offered as a saviour to others you are not related to.

He said our people are fighting Arabs dominated government. We at one-time would-be part of it, he reasoned. Then it was agreed that Alaak as an elder should go first. The rest should follow. The family sold what they had to get equipment for him. It was summertime and the area was dry.

They left Wangulei at eight am then spent days moving through Baidit, Anyidi, Aja-aguer, Mach-abol, Pibor, Wukele, Pochalla, Gilo, Zing, Gambella and Itanga.

The journey was on foot and thousand miles were trekked. He was trained as a soldier and had to fight Arabs while Ayual and Mayen had to leave few rusty books. What was left in the house of Ayual and Mayen were schoolbooks with drawings. After meeting Dad briefly at Dima in 1989, Alaak was sent to Central and Western Equatoria for operations. They went via Jebel Rad, Boma, Magoth, Kapoeta, Torit, and Pageri before they crossed the River Nile to Kajo Keji.

By mid 1990, they had intensified attacks against Arabs at outposts Yei, Tori, Raathall and Meke. They dislodged Sudan's Army Forces at Raath-all and had to retreat from enemy fire to Meke. Their commanding officers were the late Mach Guguei and Garang Bul Pageer. By September 1990, Alaak was sent to Agabe and put in charge of ammunitions and weapons. These weapons were used to liberate Mundri, Maridi, Tomburo and other small outposts by combinations of various battalions including Commandos. He met maternal Uncle Abuoi Arok Deng who was among the commanding officers in the rank of Captain before Abuoi and his colleagues proceeded to liberate most of western Equatoria. Deng joined the SPLA/M in December 1983.

Before Alaak left the village, we would smother with cow dungs and dyed our hair with urine. Then we would look like zebra. The following day, I could see him with hundreds of unaccompanied young boys. It was as emotional as the experience of not seeing loved ones again was still fresh with the news of some of the Koryom fighters who perished early on.

Around October, the other two brothers were the next batch. My name had to be replaced after mom had bitter arguments with responsible officers and having my age cited. They were told to come back after a year or so.

These boys and girls later became the Red Army organised into squads, platoons, company and the likes. They later contributed massively toward the war of liberation. The majority of them had their bones scattered over the south and east and west of Sudan. They died as heroes and heroines.

Support for the revolution was unshakeable and our chief Deng Ajang Ajak (aka Dengayang) continued to supply the movement with brave able fighters. It was sad moment, as our family meeting on who should go first was concluded. Tears were rolling on everyone's cheeks.

The family did not expect that the national cause had already reached each household. We were destined for guerrilla warfare.

Boys and girls as young as seven years old were organised into batches. The first batch went, followed shortly by second and third batch. Family was no exception. The future fighters had taken a route for bloody warfare.

Earlier on, Dad had encouraged us by saying that each generation has its own uphill battle, but he would say no one had expected this to separate families since most of the community to community fighting (communal fighting) were one offs before fighters retreated to their respective houses and camps. It was a communal or family unit sort of few hours or days assignment. It was not years of tiresome operations.

Dad would further say that in African societies, the boys were brought up and trained on fighting skills at early age. These warriors would be later organised and arranged by underage sets. Sorting them was also done to base them on the best fighters to the least. He added that what he saw on his way to Panyagor early on demonstrated the depth of the war and that everyone should have the courage to seek the outcome on battlefields. He ended with that the war was not their common stick communal fights as it would be fought with guns.

As the villagers released their children to join the rebel movements, there were sad faces across the land. Some of them were reported later to have perished in various places in Ethiopia, concentrated camps or war zones. Considered by the elders as the unlucky generation, we had to accept anything before us. The war was us and them as well and that meant everyone including Arabs died. It was a war of no retreat. The problem of the Sudan had dangerously reached the level one category.

In the South, the war was declared against the entire Arab government of Jafaar Nimeri. Then the bullets had to rain with the land littered with human bones. Men, children and women alike perished in equal numbers. It was two years after Alaak had his initiation with his age mates, that he was trained as a soldier and sent to combat around Central Equatoria. He was based there. In 1990, he was granted permission from his unit. Then we met in Bor in February 1991.

He left everything known of a first-born son to take on national duty. This call was the liberation of Sudan on a broader concept of New Sudan. The fact that Dr John Garang the leader of the movement was from Bor made his home the epicentre of massive recruitment which was later extended to the entire Southern Sudan, Nuba Mountains and Blue Nile.

CHAPTER 5

Cultural knocking of the teeth

1987 WAS THE YEAR MY BODY HORMONES were responding too and teeth were growing before my uncle Chol Young Juach was called to do his usual lower teeth knocking out practice.

In Dinka, teeth removal is a cultural tradition and widely practiced. Before Chol started his routine, he would first make sure thin wire was run through the fire to oxidise the rust. This process used to act as sterilization. Then he would insert his thinner rod below the gum and uprooting of two canines from the upper and lower with four incisors and two canines commenced. In about five minutes, it was over. Chol was quick and sweating like a cook handling hot soup on molten pan.

Teeth fillings which cost millions of people's money worldwide were going like that. I was lying on the ground before he asked me to sit upright. He showed them to me before being asked to put them in small tin. I was taken home. Game was over. I was given warm water and had to spit blood. Then hot milk. After four days, I got better, but had a little pain when drinking.

The work was voluntary. He did not charge the family anything for such talents were measured as gifts and for the benefit of the community. He was an experienced local dentist, as those without experiences would rarely miss removing one's own child jaws.

Chol was a gifted typhoid healer too. My mom used to take me to him when signs of malaria or typhoid emerged. He would ask me to sit down. Then he would call me to sit near the fire. Inside fire was a line wire. This wire must burn red before he ran it around my forehead, backs, knees and arms. These practices removed dark skin and permanent spot were left and I would look like spotted guinea fowl.

All these required one to hold their breath and an opportunity to pretend one is not deafeningly screaming. Aliet boys did not welcome cowards into their groups of girl chasing teenagers. So, one had to endure suffering.

In the summer, we would collect any available shells from snails. They would all appear in varying colours, from white, grey, dark, dark-brown, and white spotted. After this, then I would join the rest of the team members on open ground.

The likes of Gak Garang Baak, Arok Akoi Loy, Akoy Machok Biar, Dau Garang Chol, and young ones of Juach Abuoi Thon were my old childhood friends and relatives.

Figure 7 making cows and Bulls out of Clay, photo of the village boys

We would order the little ones to bring wet clay and work on moulding cows both female and male. Others would erect tiny grass symbolised huts.

One would have already imagined homesteads and set up villages with human settlements. Fat and giant men would be moulded and in imagination given names of present and past legends.

During the rainy season every year, fish either small or big would make their ways into small ponds. By October to November, when the rains regress, the water evaporates they would be exposed. Birds of prey and villagers would descend on them. By the evening, we would go home with plenty of dirty fish. Our faces would later look like the colours of hyena running on the day of the brighter moon. Some boys became sick in the process of fish exodus as running from one pond to the other became tiresome. I was not friend to dirty environments too and each year, I used to become sick. Sometimes it would become so incomprehensible to tell which type of disease one was suffering from. However, the sick would take medications made from herbs.

Whenever I became sick, Mom would ferment some known herbs. Then I would take it regardless of bitterness.

In the case of persistent sickness, village healers would be called. Others would call necromancers or seers who would later take advantage of ailments. These groups were not immune from corruptions while performing their oracles. Resorting to call for bulls as slaughter to appease the living dead became abundant.

Diseases such as severe malaria combined in the body with bacteria borne infections might end up in the category of oracles qualified for sacrificial lamb. Even after doing this, the results were sometimes disappointing.

I was known in the village for my skinny appearance. When boys with volume wrestled, I would not stand against them as weight matters if it is something to do with wrestling. I was missing this hobby a lot. It was also like a challenge to me for not doing what majority were saying to have been a practice in the village. And one day, I became disappointed, because I was prevented from joining other boys in wrestling.

When I reported that to my parents, Dad replied that wrestling is not won by the weight, but the spirit of maintaining consistent hard work. During the wrestling organised between communities around us, big boys such as Mabior Tiet and Yaak-manyang would be relied on heavily. You were required to defeat the flag bearer and his deputies. Traditional wrestling is fun based and for popularity of one's own strength. It is not by points among the players in contrast with the European football league. I would come almost second to last and all ended up in draws. Practice was essential in wrestling.

For years, my parents had hope that their ailing child would take the family name to the next village in cultural activities. But it did not happen because I was losing weight because I had measles. Then the guinea worms around my elbow and right knee were not helping this either. With local operations to remove them delayed, they were eaten up by my naturally acquired immune system. And during the day, the white stuff would emanate out with pus. It was swollen and it hurt.

The symptoms of guinea worms include inflammation and rough skins. Then blisters which appear and disappear. These symptoms happen during the incubation period. Like polio, the victims are crippled or disfigured around the ankles, knees and elbows. It would then come out having destroyed the cells. Such scary creatures are thin, long and look like tape worms.

Apart from nominating who the next shepherds to look after sheep, goats and cows would be, Dad had to also find time to join fishing expeditions which were organised by village. I would follow him to assist in removing any living fish he threw onshore from five meters deep water filled barrow pits. In addition, Dad was doing everything possible to feed us and ensuring the family was happy. He used to consider happiness as one element of peace of mind. Then I had enough of being an infant and began to herd sheep, goats and carry out fishing expeditions.

In January 1985, famine struck. Then we organised wild weed gathering. This exercise would include women, girls and boys in collecting weeds for feeding. People were hungry and starvation was severe. Those with healthy cows began to cut veins to collect the blood which was later cooked and served to children.

In the family, the responsibilities were divided. I had to do chores which were meant for girl. To help our family, I would do exactly what a girl and boy in the Aliet village could do consecutively.

When we turned up to graze our cattle in the forest, my world would outlined by two jobs: herding cattle in the day and picking dried canes for cooking. For instance, when the day replaced the nights, I would go in the morning straight to where goats and sheep spent the nights. I would make sure hyenas did not steal any at night.

And in the evenings, my mother would urge me to help her putting sorghum grains into storage. This storage was made of frames and a few stands. Hard work is rewarding. I also started to study the arts of farming. I could tell which fish and cow was pregnant and which one is about to give birth. My parents would make sure the expectant cow was always followed wherever it grazed. This is because a cow which delivered unattended could have its calf be attacked by wild animals or vultures. In some cases I would find fulfilling my obligations to the family noble. At nine am, I would then start my assigned duties with joy.

But one day there was village ritual which was organized at the Lueth homestead. I refused to attend because I had witnessed in previous rituals that there was mistreatment of children. For instance, the well-cooked meat would be put in empty calabash. Then a strong person would begin to throw it randomly everywhere and boys would scramble over it. Body injuries were common. The organisers of such activities knew the purposes of doing this. After the ritual had just been concluded, I observed a number of people eating what was dedicated to gods.

Christianity was replacing people's customs and traditional beliefs. But the spread of Christianity was not without serious condemnation from devoted believers of African beliefs and myths. For example, there was fear of offending the traditional gods. Those who had adopted biblical names such as my brother Ayual who was named after Abraham were people on the outskirts of community beliefs and norms. They were accused of having gone astray.

And by my refusal to attend the village ritual, such a sign of defiance was causing high tensions. There were people and parents who did not believe in the existence of any other better gods than what they had.

To my parents, change had come. Abraham could attend church at Wangulei and pray as he wished. He began to reject eating anything contradicting his belief. Our church in Wangulei was built of grass. It was 50 by 40 square metres. People like Rev. Joseph Kamic Deng and Rev. Alaak Kuot Duot, were among the first group to be known as Christians while living in the major cities such as Juba in the '70s. They were crusaders and spread the words of God through music and songs. English gospel songs had to be sang in Dinka. Songs like the 'Holy truth, holy, holy Almighty God (Yiic lajik Nhalicdit) sent many madly crazy.

One of our cousins lost her teeth. She fell in the assumption that Holy Spirit had possessed her. People like late Rev. Joseph Mabior Garang would later be busy at senior pastors' designated holy ground conducting baptisms for hundreds. The converts would cite allegiance to God and

and the Christian faith. They would vow not to commit adultery of any forms, not to eat ritual foods and so on. Serving the church and helping the needy were also cited and all had to agree.

Abraham and his friends from the church would organise evening prayers. I would be called to attend. Everything about life on earth and in heaven was dominating the short sermons from Pastor Joseph.

If there was shifting from the traditional, there also existed mistrust. Most of the preaching by then touched on interesting topics of freedom from evils, from suffering, gluttonousness, adultery, night orgies and so on.

Having been convinced by the life after death because in the village, elders talked of the living dead, I was not ready to convert until 1989. Offerings to local gods were no longer enough compared to the bygone days where everyone was animist. So we had that advantage. The Christianity came with cost.

For instance, married women began to neglect their family duties and spent time together doing church activities. Some would come as far as Patoor, Riakbek, Peny, Pawel, and cattle camps, Dong, Aliab and Piomahol. These activities attracted all classes of people, pundits and guests.

Such colourful events infuriated village elders. The most frustrated would even register their concerns with church authorities in vain.

Some would start accusing their wives of stocking the church with family resources. Such forms of accusations would be tabled to have the onus proven. Each December, there was enough supply of delicious food from the surrounding villagers. I did not know whether such generous free will gifts were part of villagers-church goers' contention. Also waning slowly were the powers of the heads of spiritual traditional leaders. In most churches, there were attired men and women.

Like any other organized community cultural events, the beauties, handsome and ugliest were all attracted. Temptations were also there and the churches could not control those who went against the Christian doctrines of not engaging in indecencies outside church compounds.

My conversion to Christianity was because I heard a verse in the Bible talking about the afterlife. This statement was powerful. To have longevity here and also from there. The idea of being free from the nightmares of the next immaterial world took hold of me completely. I sought guidance from the church on this transition.

I was immediately put in Sunday school and began to be taught about God. The New Testament's Gospel of Matthew Chapter Five had really assisted in spread of Christianity in Sudan. Having been translated into

Dinka dialect, our Sunday service would begin with the 'Sermon on the Mount' (King James-KJV – Mathew Chapter 5). Then Pastor Mary Achol Deng Nuer would begin her reading while I keenly listened as follows.

Both literates and semi-illiterates were busy reading Matthew Chapter Five to the end.

With the literacy rate being incredibly low, Dinka dialect and its vowels were crucial in the eradication of illiteracy too. I too benefited from the teaching. I was admired by pastors whose preaching was life reawakening. Like my parents, they cared for the sick and weak such as the under-privileged of all ages. It was undeniable that our happy childhood and upbringing was my parents' goal.

I attended church and doing other family responsibilities such as following Dad on his way to Jonglei Canal for leisure and hunting. He would grab his sticks and spears.

We would immediately walk barefooted for a distance of about twenty kilometres before we arrived where there was no settlement but a forest. Our area becomes home to thousands of antelopes during the summer dry season.

The dynamic seasonal environments and woodland and savannah grasslands encouraged animals' seasonal migration. This woodland and plains are dominated by hyparrhenia, sporobobus, pennisetum and echinochloa, combretum and balanites. This tall grassland (elephant grass) encouraged bothhunters' and large-scale migrations of giraffes, white eared kob, mongalla gazelle, bohor reedbuck, tiang, beisa oxyx and wild dogs. The wild dogs were likely left in vast plains between Jonglei State and Boma Badingilo national parks. Dad would ask me to join him in a forest hunt.

Only experienced hunters' could trace the footprints and distinguish which ones were for cows, mongalla gazelle and white eared kob. However, one must be careful because of availability of dangerous animals which came with large scale migration. Singing while in the forest was not allowed because of scaring the moving animals. Wild animals have sense of identifying approaching danger and we had to be extremely careful.

CHAPTER 6

The Hunting Expedition

Deep inside the forest, there was whispering. We stood still. In about a second, we heard someone calling a name that sounded like the Nhom-cholic equivalent of darkhead. We were stunned, because such a name was not in any way available in the village.

Then Dad grabbed my hands and we crossed the canal heading eastward in search of antelopes. I continued to wonder what it was and whether it was something to do with a stranger's physical disability or something else. Then I thought why bother when children are funny animals, who change shapes when they grew up. For no one could tell exactly who would be village cowards, geniuses or so in the future.

Notwithstanding the peace of mind the environment provided, things were changing quickly. The heaviest rains would fall from dawn to sunset. Locusts and noisy screaming foxes invaded our farms and reproduced. Rats and mouse of all sizes began to invade our house. The ecology of the area had to develop new type of soils and toads and birds dotted the ground.

Those grain eating birds when they flew would cover the sky. When they sat on our farms, no grains would be left. Sometimes they would sit on ashes left from burnt grass. There was an assumption in the village that weak ones changed into mice. This theory remains contested even today.

There were other changes brought about by modern technologies.

They included the widespread vaccination of children and distribution of clothes.

Distribution of relief foods increased and we began to taste white people processed food. Then mom would bring butter from Panyagor and it was tasty.

In our village, there were chiefs, beautiful women and cattle camps' heads. These people were responsible for the distribution of relief items. The vaccination of animals was made available. These services were not provided by the government but the agencies of the United Nations. Despite this, the majority of people were still drinking dirty water from the pools and water sources. The theory was that water had no poison.

This encouraged carelessness when one was thirsty. For example, we would drink from the same pool that frogs had laid their eggs in. The result was diarrhoea caused by contaminated water. Increased numbers of ringworm and infections as well as swollen filaments around the joints were common. Malnutrition cases of insufficient calcium left many crippled.

Years went by with many problems facing our people. For instance, the children were malnourished and sick. In the same village, we could play, dance and planted our crops. We would stay out in the rain before our parents called us inside our huts.

Then one day I saw mom with her handful of grains being thrown in the rain. This was done upon hearing thunder. There were other fears in us of tremendous tropical diseases. When any omens occurred in the village, wise persons would be consulted so that the agent behind the disease would be ascertained. Diseases were some of the bad omens.

For instance, any human settlement struck by cholera or measles would be deserted. People whose cattle were suspected to have been infested by contagious animal disease were quarantined or not allowed to mix with health cattle. This order would be imposed with impunity and even with force if one refused to submit. While on the other side of community protection, if community heads accused unknown supernatural beings of having released wrath into the villages, the infuriated villagers would over-react.

It was one message that in the village, demons would be kept by bay. One Sunday, a man came to make testimony. He began his statement and said, "For years I hardly got peace of mind and today, I want to be baptised." He rose his new acquired cross up and began to whisper to Senior Pastor Joseph of the church in Wangulei.

Then he made a proposal before the worshippers that if the pastor wanted to free him, he should also baptize his entire family, cows and goats. He concluded his testimony that he fears that deaths which had taken away his beloved children may continue to hide in the animals or hunt him down anytime.

Although the congregations were stunned by his request since baptism is limited to humans, the man was in constant fear and distressed and believed that his livestock should also be baptised for him to be free of any possible terror.

He was heard making a bond with God and took his newly acquired church name Peter. However, he could not complete the rules for him to abide as a Christian. One rule in the church was to commit to one wife and to denounce polygamy. These two items nearly caused the cancellation of group organised baptism. Other men and women free from those conditions and who did not anticipate that their plans would be completed at oath became disappointed. For instance, they had assumptions that baptism and Christian faith would take away their worries and make them free of any wrongdoing.

The pastor was left to ascertain whether the act of evil would stop sheltering man's livestock. The nightmares which the man suffered were not solely his. Some church attendants kept quiet with enormous problems although the Bible as a book of hope acted as their therapy.

Above all, the preachers were acting as psychosocial scholars. They sounded in their teaching as if they had collected each individual's life story beforehand. Sometimes, they would direct followers to the verses of Isaiah 18 saying God will punish Sudan. Then the story of Job who suffered before his blessing was granted to him by God would be used as an instrument of encouragement. Then Jesus would be praised as the redeemer of humankind, saviour and manufacturer of all medicines.

The church would pray and ask God to heal the sick. Others would come back after recoveries while the rest gave their appreciation in complete silence.

The then government was failing its responsibility to treat many curable diseases. Churches did not have enough resources to provide health services. Putting hands on the heads of ailing patients was common. It became popular that families began to take their sick ones to church. Then my prayer request was taken to church by my mother's command. My situation was deteriorating. I was attacked by an unknown disease of severe headache, fever, fatigue and unintentional weight loss, weak and sometimes I had to seek the aid of a walking stick.

By two pm, Joseph Kamich arrived carrying his bible. I was about to have chicken soup. He prayed until he sweated. He left and the next day, I began to walk slowly and take little food.

In Dinka society, many villagers were believers of animism and its practices. Church miracles emotionally moved many people to believe in Christianity as the reliance on supernatural intervention was high. One of those issues facing villagers included Sudd and its vegetation borne bacteria, germs, and viruses.

For instance, the vegetation is a sanctuary to millions of mosquitoes. The mosquitoes attacks against the village at evening hours became frequent during the rainy seasons. Preventive measures such as burning green grass and herbs, so the smoke scared the insects were not effective. Setting smoke from anything to protect us and livestock would be mobilised. By evening, fires would be lit.

The only time people had short-lived relief from the insects were on days of heavy winds and rains. Lifesaving teaching in the evening on arts of socio-economics would be disturbed by the grade one nuisance manufacturing insects. So, by eight pm, we would already be inside our huts. This hut had two windows sealed with thin sieve mesh. Apart from the door, there was only a pipe made of cow horn. It was a toilet purely meant for discharging of urine. Everyone from family members to guests used it. The rest of human wastes nourished the nearby roots of thick meadow and scrublands.

Before villagers retreated to their cow hide huts, mosquitoes would parade over the blood of innocent new babies. I would just see those smiling young ones contained inside frames made of reeds and interwoven with rawhide.

These pieces of hand crafts helped people like two moms and aunties to carry babies on heads during hundreds of miles of cattle movement. These children were so vulnerable to mosquitoes and malaria. For reasons unknown to people in the area, the nearby swamp areas were not cleared from dirty water and grass. Though the environment faced constant flooding, people had enjoyed living there for centuries. For instance, the ecosystem combined humans, livestock, weed, reptiles and all types of mammal. The millions of egg-laying insects knew nothing about the birth control. The small huts we lived in accommodated many from invisible germs to small lizards.

At night as well, we would sing songs with the belief that no amount of warrior songs were capable to keep hungry lions from raiding houses out of defence. There was also the concern of the spread of measles. In 1985, I had measles. Innocent children such as Juach Aboui and those I admired to play with could intermingle with me on playground though mom was worried about contaminating them.

To fight the disease, my parents had to properly feed me with urine adulterated cow milk to enhancement my immune system. While lying on the ground to slap and run as we played, I would watch praying mantis hunting its victims: the small flies. Such a raid was akin to the raids from the lion raids. It was well known to us that lions lay in ambush and flanks against their targets. While at night, I would hear lions sending their first warning by roaring. All animals in the forest including the cows would stand still.

Then the villagers would react by grabbing any available tools. One must only attack a lion when knowing your strong blood relatives are around. The cowards disrupted lion pursuits.

One day while I was at the Pagategee cattle camps, I saw two lions sneak in to attack herds of cattle and began to scare everyone threatening them. If they had preyed on decorated bulls, no amount of guns would have been available by then to easily scare them.

Before the civil war broke out, the girls of Aliet would respect even strangers, but with the exodus of the Sudanese, I have seen Sudanese girls and boys dropping words to nourish what the Dinka society of the older generations would regard as very disgraceful behaviours.

Village children had a good upbringing in those days. They would greet both relatives with respect, friends and opponents with great admiration. However, in our family, there was peace and great respect. That respect was also in line with Dinka customs regarded as paramount.

In those days also, there was enough trust. This trust was extended to the housing of strangers. Culture as in any other societies is considered as the core centre of any existing society.

Dinka society regards culture as the epicentre of life. That means being old in the African society came with an honorary position of wisdom and respect.

But Dad's life was cut short. He missed longevity and was taken by death. And after I had lost him, the pain of the horrors of his permanent absence began to disturb me for several months. The passing of my Dad robbed us of everything. We were orphaned at a delicate age. The other sibling got the bad news during their training in Ethiopia.

CHAPTER 7

The Chief with the Three SPLA/M Recruitment Strategy of Liberation

My dream of mastering singing skills to become a well-known singer was shattered when the SPLA began to interrupt my singing practices. Having faced the possibility of running out of fighter men, the SPLA had no choice but to look for some of us in our village.

I missed out on their registry many times. The first was in 1989 when the group of my cousin Ajak Alaak Ajak volunteered at Wangulei. This group later became the lost boys. I was among them but by luck the group was ordered to move when I was still at Pagetge cattle camp. But recruitments into SPLA ranks were serious under Commander Jok Reeng. He was an unquestionable fighter whose role in the liberation from the Sudan Army had already reached our villages.

The combinations of forced and volunteered conscriptions were conducted under court martial orders. When the recruitment was under a court martial, the best way to do is either surrender or fly. That meant anyone running away was to be shot or anything belonging to his parents confiscated until he voluntarily turned up. The soldiers did not know that I had the mathematics of equilibrium. For instance, the family had

Figure 30 Gede Historial old building, 5 April 2018, Malindi

already provided three of its sons to the movement and were already in the fighting rank and file. Among the cattle we had, none had been taken. So, I flew to the forest. I was not ready to join the Sudan People Liberation Army. Jok's soldiers followed me but they could not get me as I had crossed deep into thick thorny bushland. I spent all day in hiding to appear in the evening.

Because the chief Deng Ajang had already submitted my name, I and Mading Biar Bol took refuge in the forest adjacent to Nun Bior home. People like little Thon Garang Dau who was about six but a clever boy and few remaining disabled people were interrogated and beaten to reveal our whereabouts. Those refusing to have their children conscripted had to have cows taken. Now, it was forced recruitment. In terms of liberation contribution our family had me left to take care of my parents. I was wrong. There were families with more children taken depending on the number of grown up children present at the time.

Other cultural activities that I would have kept alive in the village were at jeopardy by joining the movement. The war just absorbed the village. The bush became unfriendly and infested with dangerous and unexploded ordinances.

Unaware of the hazards around them, children began to blow themselves up with explosives. As family head, Dad used to advise us when he was still alive not to touch anything deemed as unusual. I vividly remembered his words. This piece of advice was like a common phrase; do not enter the dark forest whether during the day or night. Without

Dad, the impacts of war were taking hold. Mom as his widow took charge of all responsibilities. She would wake up early in the morning to clear the land. Then came the nights and I would live with memory of his good deeds. I missed Dad the day hyenas broke our cowshed but Mom was brave enough to scare the hell out of the carnivores.

Then the family wealth was used to support us and the people movement. For instance, each household was providing logistics support to the rebel forces. There was a radical shift in how the war with Sudan Army Forces and its Pro-Arab militia made behavioural changes in the families. And because of war, we could not reach that time where our parents would have divided the family assets as they gracefully aged with their children. This was not happening, because my brothers had left the village. Their destination in July 1987 was another country. There was no one taking care of our cattle and Dad had decided to take me to cattle camp. By September, we left Aliet heading north. There was already a manpower gap and Dad started to train me to do what my elder brothers used to do.

So, we took off from home very early in the morning. We passed through Pawel at heart of Kongor heartland. By six pm, we decided to spend a night at a certain house as guests. We were served with Awalwal. It is a rounded food made of millet and taken with milk. Then the next day we woke up and proceeded. Here, we crossed vast bush land stretching far as Nyarweng swampland. It was only the two of us on the pathway. One could easily get lost or eaten up by animals or consumed by violent bushfire.

By three pm, we had reached the border between Twi and Nyarweng. There were too many broken dried animal bones, sticks and spears all over the place.

Then Dad confided to me that where we were passing through had communal violence caused by the dispute of water and grassing rights. Indeed, the place was looking like where a female elephant disciplined its young one by trampling it in the mud.

After we left this place nicknamed as Apangany, we trekked for another two hours. Then we arrived at the designated permanent settlement. This was the evergreen area between the two big streams. The grass was plenty for livestock. I was received by Alaak Thiong, Alaak Mangok Alaak, Dabek Mayen Ajak and Anon Garang Alaak. Boys such as the late Kuot Mangok Alaak and late Matiop Juach put cow ashes all over my body from head to toe. This was a sign of being born into cattle camp life. I was shown around the settlement which was in an oval shape. The boys who had been there the longest had to make sure that I was not afraid when handling aggressive and dangerous bulls. They would say I must know the techniques of restraining notorious cows. Cases of people being crushed or having their eyes smashed by cows were common. This ended up with owner paying compensation for the omission of their animal.

By ten pm, this exercise was ended. We then moved back to our site. I was provided with enough milk and fish. I was looked after but there was transitional shock. The borderlines between girls and boys were somehow blurred. Seeing men and women who were halfway naked was not an issue of boredom. Sometimes, I could see boys and girls sleeping closer to one another. But what saved the settlement was discipline. It was no doubt any misconduct of one person would lead to the collective flogging of the entire settlement. The least one could get was ten slashes depending on how Akoi Akecnhial as youth leader assessed the degree of misconduct. The girls also had their leaders and they received as equal a punishment as the boys. Some girls were very powerful having mastered the arts of wrestling. Boys who crossed their lines received a ruthless use of force. In fact, small boys were also trained by the big boys to misbehave. The nature of Wut was that the obstinate behaviour was encouraged.

It was an amazing experience for me personally. Wut was full of fun from group fishing to throwing wet dungs on one another's faces. The land was nature that one could urinate at will in a free soil bequeathed to one's forefathers by God. This meant I spent January to May 1988 without thinking and missing Aliet. Proteins were our diets and Thioyol, Tiep and Anyedol were our paradises.

At Anyedol, the boys were instructed to look after calves. I was supposed to join over forty boys in this exercise but I became sick with diarrhoea.

I reported it to the team leader through one of my cousins. Unknown to me was the fact someone with hate went and lied to the group that I refused to attend my duty. By four pm, the team and the calves came back on site. It was here that the boys stormed the tent I was taking a rest. I took spear to ward them off because all of them wanted to beat me for obscuring my duty. Alaak Thiong involved and the case was decided in my favour. One had to have a strong person to take care of you otherwise there was bullying where those who slept at eight thirty pm before waking up to eat or drink would find their milk mixed with dungs.

CHAPTER 8

Local Preacher Kon Ajith

It was about eight pm while looking for a lost calf when I heard a voice singing strange songs. These songs were not the ones used to invoke the big god when the village had cases of lightning strikes. The voice was from a local preacher praising the heavens and earth. His name was Kon Ajith. He was a self-proclaimed prophet.

The Moon had just arisen and the cattle camp boys were beating every container and blowing horns to welcome the phenomenon. There was a myth that the Moon comes with premonitions of varying degrees. So, we decided as boys to chase away anything bad with songs. This job was exclusively boys, and the girls were not part of it. We would sing related songs from bringing lambs for sacrifice so the blood takes away any form of witchcraft and or all evils that season. Moon provides light and enables us to milk or spot our missing cows at night but the in-ability to know what comes with the moon made many to receive it with extreme care. For instance, our people count days with the appearance of the moon. They count fifteen when the moon is on and fifteen when it disappears to arrive at months. Counting starts on the first day of it appearing.

Anything happening either good or bad that time has a significant contribution to such a natural occurrence. When the Moon is around, darkness disappeared, and we would play up to eleven pm. Then we would practice any forms of cultural activities because we were able to see ourselves. Kon had to step in here upon hearing our deafening noises.

He was not happy with naked boys running around pronouncing prejudicial judgments on the heavenly bodies. As an adult and Pastor, he knew heavenly things from morning stars to asteroids and meteoroids (known in the village as Cier Ayiol for star with bad fortune). Other stars were also helpful in knowing when the weather changes were likely to take place.

Kon had already drawn us to where he was. I was standing on the top of a hill degraded pile of old cow dung ashes. I could see Kon clearly. He was not wearing anything to assimilate himself with cattle camp lifestyle. Then he began his teaching and prophecies from Jesus to Isaiah, Ndundeng and his son Agok. These last two traditional forecasters were Nuer renowned self-proclaimers of the twentieth century. Their influenced reached as far as Twiland and many villagers used to talk about them.

Kon was of no hesitation with mixing the Bible and tradition. As newcomers in the white man's religion, we went by what he said and shouted as much as our mouths could open. He taught a short course of gospel songs. One of the songs was talking of what was prophesized in the past and strict observance of prayers from morning to evening.

The other songs referred to the prophets Isaiah, St. John and Paul's warnings to the descendants of Adam to quickly repent and those of Judah to also believe in God for the days of Judgement would have a blasting furnace and the whole world would be in anguish.

He then introduced himself as an Evangelist who had exorcised demons in the land from Mang'ok de Juet, Lierpiou, and Aleer of Nyarweng among others. He confided that he had seen God. Additionally, he said he had travelled throughout the land going from one cattle camp to the other.

Kon, as I could describe him, was a person deeply touched by Christianity. His upswing to the position of Evangelist was justified by his hard work. He asked that we disperse and to converge at church at ten am the next day. I woke up early in the morning the next day.

I collected our cow wet dungs and spread them wider in the sun for quicker drying and then rushed to the church as requested earlier. Upon entering the church, Kon was there. He had already been welcomed by the Sunday school teacher. And he then began to praise the goodness of the Christian God different from the local *Nhialic* of our ancestors. He told us for the first time that a virgin woman in a place called Israel gave birth to a baby boy called Jesus of Nazareth. No man had had intercourse with that woman and conception was by the Holy Spirit. So, the baby was God in trinity.

Then he posed a question to us as to whether such a miracle had ever happened in our village? Our answer was no, impossible. Then we were reminded of the resurrection after death. It was another new idea too. In village traditional beliefs, people think that if one dies, there is no resurrection other than joining one's kinsperson with the living dead.

They would say the body disappears but that the spirit remains to interact with the living. For instance, people who have lost loved ones mourn for several months. This would include wearing dark clothes and ropes around the necks.

To Kon, there was life beyond the grave. That life was a resurrection. When church services ended, I shared the story with my friends and everyone. Others thought I was crazy while the rest went mute.

For a few days, there were preparations among the people in readiness to choose from the two religions.

That means they must choose carefully which religion they should stick to. In ordinary traditional practices, my parents had already bitter experiences with the masters of magic spears lying to them that if child becomes sick and sacrifices are not made to local gods, the child will die. In my years with the typhoid epidemic, my parents would choose as to whether to sacrifice a lamb or hen to appease our traditional gods so I could survive.

But in that service, Kon had made it clearer to us that Christians' problems are already solved by one sacrifice not too many sacrifices. He said that is because of Jesus Christ. In many traditional ceremonies, I had witnessed in the village, they had to make many sacrifices.

The questions some audience asked were about where and when that sacrifice was made. Kon knew the answer. He urged young and old to believe in the son of God. Indeed, they believed and many were baptized. When he reminded us, that Jesus was sacrificed and whoever had burdens should come to church and be connected to the God of truths, I had the desire to connect to God. At a young delicate age, I could agree to be baptized.

I would have taken it in a different context if I had waited a few years for that connection to happen. In the church, we were taught how to pray. Then we had three religions: Christianity, Islam and traditional beliefs.

All these religions caused tension and conflict. Village were being attacked by government holy warriors for to them, being Christian was somehow against Islam. While sitting with my friends reciting "Our father who art in Heaven give us our daily bread", war planes would fly overhead. That part of gospel creed was a hope to me.

Following the famine and droughts, which killed cattle and burned crops, there was no food and praying to supernatural beings so that poor families should get their next daily bread was already in the Bible.

That among other things I came with from the service made my parents to think of the rewards for a pious life. And my father who was seriously ill decided to attend the next church's services. That was because there was hope in the church's teaching.

Imagine a situation where someone mentions to you that in our unforeseeable world, that come some day there would be a life after death. Dad took a courageous step to change from African beliefs. His step redirected the whole family toward life of hope and truth telling. I was then baptized with the name John.

After a month later, I was taught the Ten Commandments. Notwithstanding the village values of zero tolerance to vices which would disgrace the family, Christian teaching was close to our village moral values. Anything idolatry was burned and the village then became Christian free land. First in the freedom was Christianity replacing traditional beliefs. None of us in the village was for Islam. Freedom to choose between Christianity and traditional African beliefs was encouraged.

In spreading Christianity, it was people replacing the old beliefs for the new. But what they did not realize was that after their war against idolatry worships ended they would be the next victims of the extreme Islamist government because of their Christian faith.

In the Sudan, people talk about Arabs and black Africans. To many generations of Arabs born since my parents' time, being an African was associated with being a slave, something supported and endorsed by societal and political systems at the time. This was resisted and many Arab masters of slave trades were hacked to death by angry villagers.

When people talked about their children being abducted and sold into slavery to serve the Arab nomads in the north, these children were taken from the villages in the south.

The government was labelling Christians in the south of the country as infidels. For many people experiencing constant raids from Arab nomads for slavery, there was no sign that Sudan's broken social fabric would be fixed.

But there was an expansion of the extremist Islamic brotherhood in

Egypt to the Horn of Africa, and Khartoum where Mahdi had a hundred years ago branded himself as the saviour of Sudan Islam ideology passed such beliefs to his supporters. As Kon was teaching the government was ordering the killings and persecution of Christians across Sudan. These killings aimed to promote mass murder. The outcomes led to million deaths.

There was a spread of Islam but with cruelty. Then lepers in Bor were massacred. The government which should have cared for them could not bother to save them. And my next concern was that the people who had just finished their conversion into Christianity would be the next victims. Kon Ajith was kidnapped and killed by the government.

His body was cut into parts. Each piece was buried separately by Sudan army which was based in Bor town. The army was wrongly informed that even if Kon is murdered, he could resurrect and capture Bor with the help of the Holy Spirit. Such allegations justified their cruel actions against the man I knew.

His murder by government forces sent waves of panic to many surrounding villages and his followers. I was his disciple and shed tears. It was sad and shocking for some of us who had benefited from his teaching. It was Kon who suggested I should be baptized as John.

Before he submitted my proposed name to the church headquarters for baptism, Kon made sure that I had enough knowledge of the Anglican Christian doctrines.

In the villages around Bor town, people were killed. Families were scattered. Many children were taken as pillage and looting were the order of the day. Attendees to Kon's church services had grown despite it being punishable by death.

CHAPTER 9

Movement within Toch

Then we decided to move in search of green pastures. It was time we crossed to the wetland of Nuer Gaweer by following the banks of the River Nile. We found people who were friendly though I could not understand the Nuer language. Adults like Chagai Akoi Ajak were proudly blowing their big, hallowed horns. This attracted beautiful women and children.

By June, we started to move southeast toward Miirakoron and Shollo. Now, we had left the wetlands (Toch). In these places, we intermingled together with farmers. Having missed food prepared out of sorghum, we began to exchange our milk with grains. Group visits to the houses of respected chiefs were organized.

Then people would serve our groups with pumpkins and green roasted maize without anything in return. There was a proper understanding with farmers on the meaningful use of the traditional right of way. These people had cattle too but in the villages agriculture was commonly practiced.

At three pm we would go back to attend to our responsibilities at cattle camp. When it rained, we would be rained on unless we could use any available items from leather, leaves and shields as cover from heavy rains. It was endurance of hardship. When bulls started to fight, everyone would go to cheer them on. These bulls had horns sharpened by well-known brave men. Those whose bulls were cowards got bonuses in the form of bullying. This fight sometimes ended up bloody. The weak ones were worn-out while the vicious ones controlled herds of fertile females. The control of herds would force the cowards to steal.

By August, the level of water was over one meter. Dogs were swimming. Then again, we tried to escape from the water by moving our cattle to higher grounds. Our temporary settlements included cattle, people and a few dogs. The goats, sheep and weak calves were left in the villages.

The process of moving cattle from one point to the other involved the crossing of vast lands of hundreds of miles. Here, people had to cross deep, but small rivers to move to places suitable for animal and human settlement. In the villages around two Duks, we would hear raids from Murle and the Islamist government's supported warriors.

The displaced were wondering what would happen to them the next day. Then, we moved to Koorquin and then Palaaidhiem and Aper Moon Joh.

I found Jonglei Canal full to the brim. A small wooden boat was brought to ferry women and children across. I was among them. However, things did not go well and I nearly drowned after the boat capsized.

I got helping hands from my nephew Duot Atem Bul (Awerjok). Then Dad decided to move our cattle to Apeer at Kier where I spent the whole of 1988 at my nephew Wach Agoth Wach's home.

Wach was from Abek pan-kuol. In the morning, I would wake up and clean the cow mess with his daughter Akuany. We would drink cow milk and play. Then one evening, the house caught fire before it was rescued. By this year, there were not many cows left to us having thirty of them given as bride price for the marriage of my Dad's second wife Mama Nyanwut Malual Manyang.

By June, Dad picked me up at Kier and we crossed back to Aliet. Then he became sick before the family decided to take him to Ethiopia for treatment. I was left under the care of my stepmother and Uncle Kuany Alaak. He was treated in vain in Dima until when he asked to return home. I was in class with Manyok Kuot Deng and another boy called Madol. There were no girls in our class. The girls were busy fetching firewood and doing everything their parents requested them to do before being married off.

There were many children denied schooling by their parents. This was because some parents were reluctant to appreciate that anything good might come out of schools. At Wangulei, some of us had to sit under the trees. The two rusty iron sheet rooms built in 1975 through donations from sons of the area were for teachers. When the lesson started, we would use our fingers to write alphabetical letters on the ground. Poor spelling was punished by burnishing one's finger on the hardground.

Teacher was influential in persuading cattle-oriented boys to concentrate. There was no seriousness in us at all. Like bats, we were too noisy. My only clothes was a pair of shorts with holes at the bottom. When the wind blew, the badly torn parts would swing. It was after lunch break when news of my parents arriving reached me.

I immediately rushed to the Wangulei centre on its west section where Ajith Jurkuch's house was. Late Ajith was my Cousin Lueth Garang Alaak's husband. He was a devoted Christian.

Dad was nursed at home where he peacefully passed on at about six pm after few months. Our neighbour and cousin Abuoi Thon Kamich were sent to break the news to my step uncles, Garang Alaak and Kuany Alaak.

At about nine pm, he was laid to rest in his ancestral land, where his grandparents were buried. This time, his grave was dug in the middle of the cowshed. The grave was barricaded with acacia like smooth tree to represent the shrine veneer. His last word to me was to take education seriously. After we had completed the mourning period, which involved wearing ring like round and braided cow skin, *Arook*, and finally the completion of the funeral rites, I proceeded to Pakou to take care of our cattle. I oversaw many heads of cattle. I would do everything from making the ropes to the mending of torn ropes. Other activities such as cutting the tails of the young ones and feeding them was added to my timetable. When our turn came to nominate someone to act as Abiok for steer, I would go ushering hundreds of thousands of Ayual community cattle from water points to grazing zones and back. This task had to be performed by over six persons comprised of able boys to strong men.

Here, we would make whispering sounds and flogging to change the bearing of fast-moving cattle rushing to get water and grass. We would make sure by three pm depending on the distance we had travelled to return back to Pakou.

Anyone who disregarded his duty and lost cattle or had a few of them eaten up by wild animals had a fine of cows and a month of doing the same tiresome jobs. Running after cattle from eleven am to six pm could induce watery diarrhoea. We would make sure we came last after it was known that no cattle were left out.

The team had the responsibility to nurse the weak ones, the broken, wounded, or expecting ones back safely. Leaving a just delivered calf in the bush was also a crime. The laws were so tough that one must have well-built buttocks to tolerate a hundred lashes.

Imposing the regulations was the task of people like my Dad (*benywut*) and the chief Deng Ajang Ajak. This responsibility was later given to Ayual Garang Alaak. It was like a kingship sort of arrangement.

One day, I got injured around the middle of my foot. This occurred in my sleep and no one knew what peeled the skin. It was suspected to be an injury inflicted by a big bull. There was one aggressive white bull which had attacked and tossed a wrestler Dhieu Atiop at the chest sometime back. The wound was nasty. I had two other gangrenous wounds around my ankles. This added to more pain. I sustained these wounds while playing.

In June to July there were heavy rainfalls. So, I decided to take all the cows to be sheltered at home. The only form of communication available was by message senders, or by physically turning up at home.

So, I surprised my two moms. Our cow shed was big enough to accommodate thirty heads of cattle. Then we spent only few months before the attacks on villages by militants intensified.

Ayual Garang Alaak who was Dad's successor as the chief of the cattle camps came home to decide to take cattle for hiding. The militia of the White Army was killing people and taking anything they could find.

CHAPTER 10

Twi land Systematic Attack

On August 24, 1991, another unexpected bad news ran across rebel held territories. It was a coup instigated by the Nasir faction of Dr. Riek Machar and Lam Akol against the rebel leader Dr. John Garang. Attempts by Dr. Garang to send fighters to counter the advancing Nasir forces to Twi and Bor were futile. As a result, the soldiers deployed at Jonglei Canal were dislodged at Poktap. Countering the enemy at Duk was also unsuccessful. The fighting started at six am and lasted for six hours. The gun sounds began to get closer. We knew something had gone terribly wrong. Then thousands of civilians displaced from Ngok, Hol, Nyarweng and Twi arrived in big numbers. The last company of the Red Army stationed around Panyagor who had laid ambushes east of the town lost scores of their fighters before they retreated.

Mom was still at Piomahol where she had paid a visit and collected some grains given to by my aunty Nyandeng. She had to rush to us. Then she arrived at eleven am and ordered that we should quickly leave the place. The soldiers were running in disarrayed. One of the soldiers by name Akech removed a handwoundgrenade and showed me where he had kept a machine inside Akoi Loy's small pond full of water. I picked a few possessions including ten kilos of dried tobacco which was properly contained inside the sack like sieved goat skins. I also made sure the two rams were not left out though they could not make it for they had to drown. So, I left them between Pa anget and Dong. They were the food of birds of prey. There was flood all over.

Then we decided not to use the main road from Wangulei to Maar for the enemy was using it. We cut across from Aliab, Dong before we slept west of Pakeer land at eight pm. Then Manyuon mangok was shot. He could not move and was therefore left to die in the water though I heard he was picked up by stranger and taken to Toch. There was flooding and the water level was half a metre. I cried upon hearing that a boy we used to play was no longer with me.

His injury was not life threatening. It was around the leg but the bone had no fracture. Possessed and overwhelmed by what would happen to Manyuon and I could not sleep.

Then we were woken up by huge sounds of fish wrangling over carcases and dead bodies. Manyuon was lucky; he got picked up by a good citizen and taken to the impenetrable island at River Nile.

We then rolled our mosquito nets and moved. By five am, we had travelled more than several kilometres and were already at border of Twi and Bor. The path was narrow and beside it was nothing but jungle. It was the rainy season and we had to pull through the mud. The sound of guns scared me. And sometimes we had to take few minutes rest under the trees. It was a catastrophe as thousands of people who knew nothing but a war crossed vast flooded area to Bor. We used the safest and shortest route to escape from Aliet, Aliab near Pa-nget, Dong, and between Palaau and Toch. We finally entered the dirty muddy road south of Maar leaving it behind. Tired after pushing through floodwater, we slept by the side of the dyke. Then the next day, we arrived at Jalle and via Makol Cuei between ten and eleven am. We slept on the roadside.

The whole strategy of such targeted violence was a counterinsurgency calculated to hit on Dr. John Garang's birthday. Left behind was Atemyath shrine and everything of important heritage of Garang hometown's Wangulei. Now, the violence was not the Arabs against us, it was brother against brother.

I survived the first attack against our village at seven am because SPLA soldiers comprising of Nubians had rescued us and our cows. We were stranded under the water having left the grass houses in fear of being burnt alive should it get hit with missiles.

Then once flying bullets hit Baak Juach Alaak's house. It was in flame. Aguer Bul and I could hear the smelling of burning chickens and goats after the family had luckily escaped. The horrors of the civil war came to us in form of looting, child abduction, mass killing and extra judicial executions. Girls and women were abducted and taken. Agau Garang was abducted between Bor and Maar. I met her children two decades later after abductees had returned her. Some of families who made it safely to Twiland's forest died of starvation in hiding.

We had no option but to escape which was a marathon. We did not know who the next victims would be. Then a certain oracle frequently cited by elders came about. For instance, the elders used to say that the Nuer prophet Ngundeng had predicted that the short hair generation would scatter all over the world.

While walking, I was crossing, poorly dressed and barefooted. It was raining heavily. There were many lost and abandoned children in the group.

With parents caught by fire while in different zones, many of the unaccompanied minors had cut through cobra infested forests. Tired and hungry, some would sleep on the roadside watching as people moved. Others would lose sight of us. Only God was providing the defence as our line of defence was so vulnerable without a gun or spears. But one thing which kept us going was a lack of fear. It was accepted that nature would do its own selection. When there is commotion, we first ascertained where the sound was coming from before the command to start running was whispered. One could easily get dropped from the group. Then we would get fire produced out of friction to warm ourselves and boiled grains. Heavenly bodies such as stars were relied on when we lost our direction at night. One of our friends John got lost. He was never found. No one knows whether he was eaten by animals or died of starvation or thirst.

There were no watches. This meant we had to monitor the change in the stars. Our footprints would terrify jungle squirrels. In the day, our eyes would look reddish like that of black pigeons. We arrived at Mabiorgon. It was flowing with water at a fast speed. The wounded, sick and children young as eight had to be lifted up. We crossed successfully before we arrived at Alilir and Akuai Deng and Mach Deng. On the way we met several soldiers sent by the orders of the leadership of the SPLA to defend the land and the civilians. This was reinforcement sent by the SPLA under Dr. Garang's commanders. People like artilleryman Angui showed their guns in readiness to fight. Behind us at our back, the fighting ensued and lasted for three days.

By the time we arrived at Yomchir, the late Ajak Mayen Ajak, a well-trained soldier since Koryom arrived with his men. He instructed that we pack our stuff and moved quickly.

They were carrying the wounded and he was injured around his knee though it was not a bullet but from an AK47 dust cover (receiver cover).

The only BM-21 mounted with 122 mm multiple rocket launcher and basically fired 40-rockets was blown up. Other mounted machines got stuck in the muds and had to be abandoned. Few of 12.7mm machine guns had to be carried by strongmen taken from us. It was a defeat with casualties on both sides. The brave resistance the soldiers put up enabled thousands of civilians to escape to the town of Bor. Those caught amid gunfire were murdered.

The few soldiers who joined us on the way when asked could not talk. They would order that we moved quickly. Then I became sick before I arrived at Mach Deng. I took bitter leaves from a local tree we knew as medicine. I got better and then I moved. A few of our cows Ayual Garang managed to sneak to Komerek were taken with some used by soldiers as food.

Soldiers have one thing in common: they rarely discussed issues of military nature when they are still in operations.

We spent several hours before we arrived in Bor town at ten am. Having heard the rumours of the impending enemy, the Bor population were busy packing whatever belongings they could carry. There was looting of relief food and other items. Then I took biscuits and plastic sheets which we later used on the way. I had to make sure I took enough oral salts as well. By three pm, we left Bor before we arrived at Malek. We rested by clearing the bushes under the trees and slept. We spent about two days before the news reached that the enemy had already set up a flank east of Bor town and Malek. By five pm the next day, we left passing through Panpandiar and then to Pariak in the evening. Then we took a rest by River Nile bank at Pariak. At dawn, there were gunshots. This was the enemy firing at everyone. Many people drowned. Then we first ran in disarray before my Maternal Uncle Malek Arok Deng grabbed my hands, Konbin and that of Manyuon.

Women who were part of us had to be encouraged to stay in the middle. Then we ran twisting, but we had to be in columns. First we ran toward the River, then we changed course to west of River Nile and cut through thick acacia trees and shrubs. Gun sounds ensued and we lost uncle Garang Alaak Ayual.

Having been attacked from three directions, uncle Garang Alaak who limped on one leg having had an accident in the '80s in Juba could not escape. I assumed he might not have been able to run and was murdered. Ever since, his body has not been found. That mean his bones lie somewhere unburied. Killed for unjustifiable reason, his spirit would one day ask for those who destroyed his life. He was a peasant and a farmer. Like most of the villagers those days, he would go to Wangulei and paid his tax of family members and livestock. He would not ask tax collectors why his grains and cows were to be taxed when he had never tasted services provided by the government since their age set of Makol.

Instead his murderers did not give his body a dignified burial: he was left to the birds of prey. And vultures which had tasted human flesh dotted the land as they followed every movement of the distressed civilians escaping insecurity.

By evening we had made it to Gwojoadung and the next day to Jemeza. We slept over with a few SPLA fighters. They were patrolling until morning. A company which had laid ambush and made my eyes widened while bypassing them as I thought they were enemy and would be dead. They whispered that we should move quickly. In less than two minutes there was an ensuing running gun battle. The fighters abated enemy pursuing us. Then we were able to breathe as our attackers retreated to Pariak for reorganisation. By twelve pm, we were still in the bush and we had to decide to move back to the main road connecting Gwojoadung with Jemeza.

Then we met many people moving with young children and elderly. We ordered that people should rest for one hour at Poko.

Also, we assumed that without sounds of the guns, the enemy was far from us. Then we collected firewood and boiled maize. By two pm, we were attacked and Konbin grabbed the cooking pot. We poured water out from the pan and asked Konbin to put the pot on his head. We used old dirty cloths to assist him while the pot got cold.

They were not well cooked though we ate them as we moved. By seven pm, we arrived at Bala pool. Then we found Mundri people crossing the River Nile to Terekeka.

We were served with ground nuts. I sat around the fire with two boys who lost their parents and found themselves moving with strangers. In our team, we had stepmother Nyanwut, Ajak Atem, late Arok Abuoi Arok and mom who got lost in the bush.

Having been surrounded by the enemy, they spent three weeks hiding inside undergrowth. At Bala pool, we slept until five am before being told that the Sudan Army forces had sent fighters to attack us on the way. So, we took off and by ten am on Mangokolang road, we were hit several times by aerial fire from a warplane. No one was hurt, as we had to take cover in trenches. By eight pm we arrived at Mangokolang and slept there. By six am, we had trekked several miles and we could see Mogri Mountains at the horizons. It was my first time to see a mountain because our village is flat lowland. Then I whispered to Konbin how amazing it was that Mundari people could built high ground from the sands.

By afternoon, we had arrived at Pawuoi where we spent a day. At four pm, there were incursions from the Sudan pro military of Murle. They breached the line of defence at the small bridge (Khor) south of Pawuoi and one SPLA soldier was killed. Machine guns were used, and they had to run away. By ten am, we arrived at Mongalla. We spent few days waiting to receive United Nations food distributions. This was interrupted by constant air bombardment. I, grandma Abuk Deng Kuir and Bior Abuoi were luckily missed by shrapnel.

Four bombs landed and three did not explode. The one which exploded killed three people and injured Mary Achol Manyiel, her daughter and son. Achol lost her limb and her son Majaang lost a hand. The daughter had her upper lip slashed out. The Sudan Army together with its pro militants were committing unspeakable atrocities against civilians.

It was already end of November 1991. By seven am the next day, we left Mongalla with the intention to go to the SPLA held town of Torit. While loitering in the dusty and dirty town, I bumped into Alaak Atem (Adeen-wut) and my nephew Atem Diing Atem at different times. Both were Koryom fighters who were stationed to defend the town from the enemy. I was taken by Alaak to their camp located southeast of Mongalla town. They were posted there to provide reinforcement to the town in case the Sudan Army Forces (SAF) entered the town from the Nile River. I was provided with goat meat mixed with maize by Alaak. By three pm, he asked that I should go. It was shocking news to me that he was killed around Juba in 1992 while defending the land. He is one of the unrecognised heroes of the armed struggle against Arabs. Atem whom I met in Labone in 1994 was murdered by his colleague about seven years later. Then my hope of spending more time with them was shattered.

Having left Mongalla, we rested at Gori and then crossed to river Nyigera where we slept and then Bilnyang where we spent a day.

There was news of the enemy having been defeated and Borland being liberated as of 21 November 1991. Notable commanders such as Salva Kiir Mayardit, Bior Ajang Duot, Kuol Manyang Juuk and among others of the high command who were around Juba decided that the homeland must be defended from the Arabs and its allied militants. By December 1991, we had already made it back to Pawuoi as the daily bombardments of Mongalla were unbearable. We spent Christmas Eve under the trees drinking dirty water from the small pools and eating plain maize and wild fruits and leaves.

We were joined by my two moms having been told by strangers that two women wearing wooden crosses were found in the bush between Jemeza and Pawuoi. They had with them two children.

Then we had to wait for them until we reunited at the outskirts of Mongalla. We celebrated and praised heaven together. Now, I had parents instead of being orphaned having lost Dad year ago. By the riverbank, I decided to sell tobacco I brought with me from Aliet. It was bright sunny day at 3.30pm and I unintentionally stood in the midst of the road and urinated. I did not know that it would irritate one of the officers standing under the shade of big mango tree.

Mongalla was dirty town with no nasty latrines or none. With remnants, and unexploded ordinances and ruined of war laid everywhere, I did not realise I was committing a crime. The officer came from behind, grabbed me on a collar and released a powerful kick at my butt. I ran away in pain. At the cattle camp I had left back then, there was no designated place for urines. Now, things were different and civilization was moving in and I had learned my lesson in hard way.

By January, we moved to Malek town. This decision was based on the order of my uncle Malek. His name was derived from the town. He said they had lived there in midst 1964 when their family had escaped the devastating floods.

From there Mom and Malek went back to assess the situations and conditions of our relatives left stranded at Twiland.

Surprisingly, 9 January was a jovial day. By ten thirty pm, while sleeping inside the white mosquito net tethered to four small poles, there was a voice. The voice of my brother Alaak. It had been four years since he left our village in early 1987. I did not expect him for we had not been in contact for years. But Mom recognised his voice. She jumped out of the net and they hugged. He came along with Aunty Nyandeng Kuir Aleu who crossed from Western Equatoria to Torit. The two on permission had left Equatoria to assess our condition having been told that Sudan Army Forces would attack Bor during the dried season. They found us in appalling and deplorable conditions as many children were dying and burials were made every day.

This was because Malek was hit by starvation and malnutrition. There was an outbreak of cholera and other water borne diseases. There were many human dried bones and bodies lying everywhere. Those were people killed during the enormous militant attacks and there was no time to bury them.

At the riverbank, I found the bodies of more than six people in one place, some with ropes around their necks. They were difficult to recognize and I had to assume that those were the bodies of our relatives of Garang and Deng Guer Deng. It was in midst 1990 when I spent a good time with the late Deng only for him to be murdered in 1991. He was sent back from the Zing training camp on orders of Commander William Nyuon Bany.

He was lovely, funny, and acted crazy sometimes. I heard he had to fight fearful William Nyuon at the parade at Zing and took the commander's commando cap. Only a man with really big testicles could act like that without receiving a firing squad's orders.

In those days, senior commanders of the movement were using the Niccolo Machiavelli's theory of better to be feared than to be loved.

Late Deng had to survive his ordeals with comrade William only to die in the hands of brothers (Nasir Faction) whom they trained with to fight the common enemy.

Because Alaak and Nyandeng had money, we were able to buy fish from fishermen and our health improved. There were enough biscuits to boost our energy. We spent two weeks together. Then they went to the area commander, Chagai Atem Biar, who was based in Bor town. CDR knew Uncle Abuoi Arok very well and he as a result provided us with a boat nicknamed as Nyan-juan (for a small human eater who used to terror villages in thea mid '80s). There was theory in the whole of Borland that the Nyan-juan were human eaters of a dwarf nature. It considered as the girl of underwear because people who said they had encountered the monster were saying it wore underpants all the time.

On 29 January 1992, the boat (Nyan-juan) docked at Malek. Then we went on board and followed the river southward to Pariak. This was the same Pariak where we had lost many people and I did not sleep as I thought of the bloody day three months before. By six am, we woke up and sailed further south before we arrived at Gulyar at Awerial County. We spent three days and the SPLA lorry which was supposed to take us was full of ammunitions and soldiers so an extension with bamboo had to be made. The news of people falling off the top of those military vehicles was rife with some severely injured or crashed to death.

Then Alaak decided that we should support ourselves with hunting. Every morning the two of us would cut through the forest. In three weeks, we were at Gulyar and we had killed over four big antelopes. I nearly got crushed by group of angry baboons while running after wounded wild zebra. Alaak cocked his gun and the baboons escaped. To ensure durability, we smoked the meat and packed them in sacks.

CHAPTER 11

Gulyar Crossing

By mid March Alaak went to the SPLA outpost near Terekeka where he had a radio conversation with our maternal uncle Abuoi.

As an officer who left his university studies at Juba in 1983, he was deployed around the Bo Bridge waiting for the arrival of the Sudan Army forces sent to liberate Western Equatoria.

We had just crossed the villages deserted by the Aliab people and we met about forty people displaced from their village three days before. They told us that after the Nile River lay before us bushes and woodlands. In the woodlands were herds of buffaloes and wildebeests. There was also the issue of anti-personnel landmines. We were able to get some ground nut and wild fruits to give us strength. To make it to Tali we heard stories of ferocious fighting between the rebels and government forces two weeks before we arrived. During the night, there were bushfires caused by unknown sources. It extended from west of Tali and displaced rats and giant snakes. Having sensed the danger, we took off eastwards. Ahead of us were terrains and valleys which constituted watercourses

Most of the battles fought around major cities claimed an army of lives. It is impossible to bring them back but one senior officer I met in the Army Headquarter was saying that building a memorial for their names would be the only way to remember them. While I read *Emma War: Betrayal, love and death in the Sudan*, I found something fascinating about the Sudanese. It was one of the Europeans who previously worked in the Sudan saying that Sudanese words are always sweet and the problem with their words is that they vanish within a second. When you come back you rarely find the same thing. Even what the general was saying would be different from a description on a memorial.

They would explore the surrounding environment before the next team request was made using the sounds of fingers cracking. The sound could be easily confused with those of the birds such as macaws and cockatoos and African Grey Parrots. Amassing in big numbers was risky and attracted enemy air reconnaissance.

However, there was sometimes a big number of people asked to spread across and move over a distance while also taking into account concealment and snake navigation. It was process involving placing green tree branches on your head as Antonov12 and An-26 flew overhead.

The job they were doing was also risky, killing unarmed civilians and terrorising everyone living organism on the grounds was not a job description of honour. The war had horrors and benefits. Having been displaced and having left everything behind, I remembered my arrival at an SPLA outpost, which was a privilege. Laying in ruins on the grass was a Sudan Air Force Bomber destroyed by SPLA missiles. It was an opportunity. I spotted it on the ground just few yards from the muddy road. Immediately, I proceeded and did full inspection of the wings, wreckage engine, and tail and fuselage tank. The crew were said to have been sent to Deng Nhial.

Deng was considered by majority of Southern Sudanese as the divine of the oppressed people. They believe if he is in good place in heaven, his spirit and that of other forefathers could adjudicate over their matters with oppressors. The crew were of Arab origin and mostly from Middle East countries and did not survive the crashed.

In war, it is millions of unarmed civilians who become the victims. Forced from their homes, scattered, maimed, killed or tortured, the survival of over six million Southerners displaced to neighbouring countries. In the middle of nowhere, we found a man. He noted from the hanging diaphragm that I was hungry. He had with him sweet wild roots and gave me two. I shared with others to give us energy. He also gave me the tiny pink leaves of the thyme plant. He added that, it was medicine for stomach discomfort. In local Arabic, he said that the herb cures parasitic worm infections and skin disorders and intestinal gas.

As a result, I took it, packed inside 10-kilogram USAID marked woven polypropylene bag. I took it during looting of UN food items at Mongalla. The few civilians we found in the villages were so worried for their safety. They stopped farming because of the insecurity. The starvation was imminent and most of them were living in undesirable conditions.

Starvation and related deaths were complicated by the restrictions imposed by Sudan government on the United Nations relief supplies to the most affected areas. By two pm, facing the direction of the sun, we left the village of Bullu.

Those who were left behind because they could not walk properly received unpardonable blame from the first group. Words like "Why are you walking like a tortoise: do you think we care to be waiting for you?" would be thrown at them. In return, they would look to our eyes and say nothing. One had to be careful blaming someone who was tired and had a gun. Frustration could be turned against you in a gun shot.

We had to wait for the elderly women wobbling with the help of sticks. And doing that meant we would have to wait for thirty minutes or so and started again when the leader ordered the move. If no one was missing, the leader would tell us to move.

And if there was one carrying a gun, he would be required to alert the group lagging by firing the shots into the air. Because many of the paths were into the forests, I remember the day when I could not move properly, because I had sores on my feet and around the knees. With my water container in my shoulder and a few possessions wrapped in the plastic sheet, we started the journey together but after two hours after we left Tali town, I began to fall behind. No one was with me and I tried to run but was still falling behind. An hour later, all people in the group were not in my sights. I continued moving but I could not make it as I thought. What was even coming to my mind was to take rest. But the night was near and without some people around me it would be dangerous.

I increased the pace and when I arrived, everyone was laughing when I told them I was walking as fast as I could. They said I was not walking but running.

While I running there were sweats rolling from the head all the way to both feet. My joints were rattling and exhausted. Somewhere in the middle I had found two paths, I had to critically look for which one had footprints. Then I found a cross marked in the middle of the path. While at in the forest, we had agreed to use the sign as a no go sign for the path.

We had four or five women in our group. They were carrying babies on their shoulders and water containers full of water. They lagged behind for ten minutes and when they arrived, they alerted us that the government forces were rumoured to be approaching. And that we must be mindful of our safety. The next day, we passed through dotted villages located in beautiful landscapes.

The people there were cooperative and supportive. Some chiefs for example would order his people to collect food for us.

Then we would sit under the tree and shared what has been provided to us. The generosity of Equatoria villagers was impressive. Despite such commitment to give, I found many were on the verge of dire poverty. We arrived at a small concentrated settlement before Tali and then accommodated ourselves at an rusty iron sheet house. The upper part of the house was burned to scorch by missile and the ruins inside had overcooked animal bones.

Then two soldiers came to us and informed us that there was frequent fighting around areas south south east (SSE) of Mundri. This was between the rebels and the government over the control of some outposts.

The house was stinking because the bats had laid many drops and urine inside it. Other adjacent houses had iron sheets hanging torn by shrapnel. He left only to come back carrying a quarter bag of beans. That was a relief to starving children. Everyone was excited. Then we divided ourselves for those to collect firewood, water and those to light the fire out of dried grass. We poured enough beans on pans, and after two hours, it was ready.

CHAPTER 12

The power of courage in a trying time

WE MANAGED TO TREK several miles because of our courage. Also, at first I did not take it seriously enough to be completely consumed by fear. Socrates in his philosophical reasoning had it that the uncontrollable fear amounted to defeat and one could easily get hurt. We were living like guerrillas; we would pass through thick grassland without taking into considerations antipersonnel landmines. There were so many left landmines either for heavy vehicles or humans implanted across areas of Southern Sudan and we would without fear penetrated through the bushes to either collect firewood or get grass. Unfortunately, there was a loss of lives to hunger, famine and disease.

In such circumstances, we would resort to anything to survive including birds. Small rodents like squirrels, big rats and dried meat from the bush constituted the tasty soups of those bygone bad days. I spent years without knowing that there was somewhere processed clean water. Whether raining or not, we would make sure our containers had enough water fetched directly from the river or streams. Stories of people dehydrated to death were common. Those who lost loved ones in such circumstances had in their minds the indirect consequences of the war. Hence, they became civilians bearing the brand of the armed conflict.

To survive, we had to endure the suffering and considered that we shall live so long bullets have not cut our lives short. But the people we met on most of our journeys were so cooperative. Before we left any area, we had to make sure we had enough information on the positions of the enemy. I could remember vividly when a certain sultan (chief) around Yei walked us along the forest to divert the enemy line. He was helpful as he identified areas infested with landmines. As we moved along, we had one goal in common; ensuring every one of our group had arrived safely and to care for sick ones. In case we found path intersections, we made a note on how to get through with a heap of sand or with a tie bundle of grass.

Only those animals considered as taboo like hyena were not in our eating menu. Our people had a belief that eating the meat of hyena makes someone boo like hyena in addition to thinking like a hyena. It was not a good idea for one to be considering.

Most of the areas we passed through had no dwelling settlements. For instance, a vast area between Lake States and Mundri remained the home of giant elephants, lions and baboons, buffaloes and all kind of savannah born animals. Before we reached Lui we found a pride of lions feeding on buffalo. In our group, we had only two guns. These were not enough to provoke hungry lions with bloodied mouths. We moved faster than we thought.

In the village those days, we were advised that the unjustifiable disturbances of a lion could lead to an unforgettable response from king of jungle. Our nephew who provoked a lion at Aliab in 1985 to attack a sleeping lion survived the jaw and claws because he was ten. In response to his spear, the lion slapped him with his paw and knocked him down. All the children watching this were horrified. Some stood waiting to see his throat being cut and the rest fled. The lion under no threat left him unharmed. He had only bruises and no nasty wounds. The relatives who ran to his rescue found him in total confusion. He was taken home and cared for for a few days. There are theories in the village that debated whether it was an unintentional or intentional attack. He was lucky. Having left the lions behind without any form of disturbance, I thought of what would have been our group had lions decided to terrorise our line of movement.

It would be like it was when we heard from the two villagers we found on the way to Tali where a man and his wife had spent four days on the top of a giant tree. They were chased there by the lions and had quickly sought refuge. The villagers had mobilised all forms of tools and they were finally rescued.

There was the time we found unspecified bones and we would not bother whether those were from humans or wild animals. In the midst of the forest, only strange sounds emanated from the thick and quiet land of no immediate rescue.

When we got exhausted, we would sit under the trees. I took a little water and kept the rest. We were taught that one should not rush in pouring out the water he had in the container even if one finds plenty of water in a river unless there is no pending enemy. If the enemy attacks, and container is empty, there would be no time to fetch the water. They would say if enemy attack, we should still run with the water in the container.

In those days, there were places where we could go for 72 hours without water. Water sources were to be marked and the capacity of water one had to carry would be taken into journey assessment consideration. Others have died because of thirst. I remembered being informed that some of soldiers who captured Yei and who later recruited us into their rank in Narus died of thirst between Pibor and Kapeota. Mathiang Diing who trekked several miles came back to Narus with swollen feet. He narrated that their guns were recovered in the desert. They left their post illegally with the intention to go back to their villages. Unfortunately, they were doing the illegal possibility during the summer. Vast areas of South Sudan become dangerous during the dry seasons as water retreats from the Nile banks.

While walking the forest, there was time when I felt like this ordeal was not going to end and had little hope life is possible. Those who made the decision to be abandoned in no man's land we encouraged to keep the fight. And that mean we had to soldier on and push through difficulties of having to live like woodpecker. It was survival from the ordeal.

Then we had two boys who roasted the raw cassavas they took from someone's farm on the way to Tali and were eating the poisonous cassava. The poison is in tapioca which makes up the starchy pulp roots of this tropical plant.

By midnight, we were woken up by suffocation. Vomiting was induced. Had the village woman who specialised in cassava farming not attended to them, the boys' next day would have been a permanent goodbye.

Sometimes there would be a fight between the SPLA and Sudan Army forces during the cassava searching mission. Everyone was starving even the Sudan Army forces as the politicians in Khartoum forgot those they sent to frontlines.

Before Yei River, we were joined by two SPLA soldiers. These made the number of guns in possession of trained soldiers to five. Sometimes I would take one when the owner had gone for premeditated call of nature. Then I saw warthog between two short shrubs. I whispered to the soldier but he could not hear me. By time he rushed to his gun, the warthog was already dead having been gunned down by the other soldier sitting about a hundred yards away. We lit the fire while the rest were busy distributing the meat among the members. Then left after our lunch because we were concerned about the security of the place.

With the Mundri direction approaching, we felt like it was not like other days where we were relying on stream water. Mundri was not like jungle because it had rusty hand pumps still capable of ejecting water from underground. After about three minutes, a middle aged man came across. He greeted us and said he wanted to drink since the tap was on the midway and benefited everyone. He knelt down and fetched the water with his hand palms as I pumped. Then we asked him to show as the direction and remaining distance. Without hesitation, he gave us the precise direction and number of hours to take. We took off, and by evening, we had already arrived in small dusty town few miles from Tali. Tali was a commercial centre where businesspeople and cattle traders from diverse backgrounds met.

Then we spotted a rusty built concrete house. Hanging on the roof were bats and moving spiders. I entered inside and cleaned all the drops of any sort of faeces of living things.

Outside were broken twigs of a mango tree which I later used to cook lentils. At about seven pm, an SPLA officer known as Rock and in charge of the area came. He addressed us and promised that an escort would be provided to us in the morning. I could ascertain his authority from the way his bodyguards surrounded our building. He ordered that we shall be provided with logistical support. In those days, anyone in charge of UN foods was so respected, and acted bossy like the Governor of the Central Bank managing money in isolated and poor territory in the middle of many ungoverned territories.

Rock was a nice person. However, he had the proverbial Dinka description of a snake with a tick on the head. Our people used to say those days that the proudest is the form of how a snake with head infested by ticks act. In the morning, Rock's orders worked. We were provided with enough soldiers to provide protection not only as unarmed civilians but to ensure our belonging do not become the property of pathway robbers and violent thugs. He was posted there to counter any form of enemy flank on Mundri. Before we left, he assured us that no one would forcefully recruit us on the way or take some of the few guns in possession of group members. Such order though oral was enough to override anything written on the papers.

In our group, we had some smokers who could roll tobacco with leaves. Some people began to misbehave by tearing up the bible. Also, there were other rumours spreading that the government forces got lost in the forest. Such statements were common but sometimes untrue. There was honest human intelligence and dedicated intelligence officers in the rebel movement. Some could even penetrate the enemy line and report back the hard-proven details of their activities. Also, people were tired of war, and if they decided not to fight, the briefing to commander would include lies not limited to the total disappearance of enemy from their sight. Rock, had struggle embedded in his whole body: he could not believe and had to order men to provide security wider to the vast land.

By the language people spoke, we knew we had crossed the Bahr el Ghazal to the Equatoria region.

There was no car available for us and by five thirty am, we left Gulyar passing through Khali, Bengeya, Tali, Jakari, Kasiko, Kakamabi and Bitti. To avoid the enemy, we had to cut through the middle of the forest before we approached Yei River just north east of our direction.

The people we found at Aliab-land and cattle camps and Moroland were friendly. They provided us with food and milk free of charge. Logistics support of people for either SPLA or anyone perceived as future recruits were so common among the populous in the south. This was because the movement had asked for cooperation with the people in return for better treatment from them.

At Paap Anok Nyengir, we found a group of starving villagers mainly women and children in fishing exploration. Then we begged them for spears and we managed to kill enough fish to feed ourselves. The villagers were amused since some people with guns would just wait to loot from starving women. Alaak as a team leader was praised and I could hear something like more blessings being thrown at him. Guns in those days was power and one could prey on victims' food though looting was discouraged by the top commanders.

After three days of walking shoeless, we crossed road people who were saying we were between Yei and Mundri road. Before water came out of that hand pump tap, two people must pump it. There was not enough water, and when I jumped in, the tap hand broke.

That was bad news to the villagers living there. The little drops of water from the hole were smothered with a swarm of bees. It became unsafe to be a thirst-quencher. The villagers got upset and Dut who was standing near the tape was hit with big stick. They yelled at him that he had brought them his bad omen.

We intervened and the group fight was solved. Without no choice, we capped the bee infested water and used it for cooking the little finger millet we had in our possessions. To fight ourselves over unintentional mistake was how we convinced Dut. In those bygone days, we were told as children, that those who are weak hearted miss great opportunities ahead of them. Though hungry, malnourished and weak, we desisted from discussing the issue of what to do with what we left back in the village. It was already the rainy season by the time we arrived Mundri. The Western Equatoria as I have known it could feed the entire East Africa if its lands were properly utilised.

Inside, Mundri nice-looking mango fruits and apples lie on the ground in the forest. These were wild mango trees without ownership and could feed any moving stranger to the fullest. We grabbed a lot of them and we ate before I put some inside my bag because I did not know the town was home to delicious fruits.

A yard away from us was Mundri Bridge which we later crossed at nine am from underneath since part of it had been ripped apart by landmines. We were concerning about landmines and unexploded ordinances in the sand. We were directed to the commander in charge of the town house.

Near the commander's grass-built houses, hovered the flag of SPLM/A which comprised of black, red, green, white, blue and yellow.

I was told the Sudan flag was lowered after the enemy fled away. The sign of victory was made even stronger. The colour of our flag was different from the flag of Sudan government. The administration was also varied with no strict application of sharia penal code in the rebel held towns, such as Mundri.

Whenever the SPLM /A captured a town, administration would set up a civil authority to run the town. That was the nature of Mundri by the time we arrived. People, who fled to Congo, began to come back.

Daily, I would watch SPLA/M flag being raised in the morning to be lowered in the evening at five pm. The flute would first be blown to alert everyone for a standby. Such honour paid to the flag would include both civilians and the SPLA soldiers guiding the flag from night hooligans.

Unintended ignorance to the raising and lowering of the flag would attract punishment, but not imprisonment. A slap on the face would be enough and enforcement of obedience to what belonged to the people. The flag was the symbol of the new Sudan's glory and liberation redemption.

A few of the goats near the flag stood when seeing the military parade and matching. Apart from the usual activities of SPLA organisation, the town's security situation was stable. Because of its location, the town was not the subject of constant air raids. Some of the few schools in the town were made of substandard of concrete houses between beautiful mosques. The contrast between mosques and churches in the Sudan were visibly portraying a religiously torn and divided nation.

One day in Maridi, I attended the church after more than one year. I had to wake up early in the morning because most of Sunday services were planned to finish before eight am. This was for safety reasons as it was decided that if the Antonov comes, it would find everyone already dispersed. Maridi is deeply inside Western Equatoria hundreds of miles from former Zaire. The warplanes would come from nowhere and conduct raids. Then people familiar with Sudan Air Force Bases would say the planes had bases in Juba, Wau and in Darfur.

The town which was recovering from war had people from diverse backgrounds including the SPLA soldiers. The town had many people from different ethnicities and most of the Sunday services had to be conducted in Arabic. I would make sure I left home early before ushers would take me to where the Sunday schools sat. By six thirty am, the service would start with songs followed by preaching in Arabic and I would go home having learned nothing but the word Amen.

Then I began to understand little Arabic but not to the extent of talking with confidence. While still in the church compound, I heard the bell being rung in the nearby outpost. The bell was an alert to everyone to seek shelter from the warplane. Everyone in the church ran in disarray. We were almost crashing into the little ones.

I then looked around, and I saw everyone was running even the pastor, Lado. I immediately, I took cover in an underground bunker. I only came out after the warplane had made two turns and left. Luckily, we were not bombed that afternoon.

By twelve am, the same plane came back. It dropped several bombs a hundred yards from the police and near Deng Ajang Awai house. He blew up the house of a certain sergeant who was accused of negligence and wasting the fuel on the day of the 16 May celebration and was arrested. I was woken up by the sound, but I thought it was thunder.

By six am, the news reached us that it was a night raid in Sudan Air Forces in anticipation of the SPLA preparation to attack Juba. The bombardment might have been directed by someone who had lived in Maridi before.

It was accurate and missed the other two tanks by few metres. The sergeant was lucky because his rusty iron sheet house was destroyed in his absence. I rushed there though I was looking for groundnut embedded underneath abandoned farm. Then it became talk of the town that someone in the town might have given information of troop movements to the enemy.

CHAPTER 13

The smoked bush meat

I WAS INTRODUCED TO SMOKED BUSH-MEAT in Western Equatoria. It is delicious when taken with okra. In time of endurance and toughening conditions, anything to keep on soldiering is permissible. Then I had a taste of warthog's dried meat. It was tasteless even if combined with groundnut.

The bush meats were different from the taste of water lilies. Lilies have no nutritional value. Hallowed as such, its flowers made it looks attractive as if it could save any life. People were eating it out of no choice. In our minds, we had that no matter how long and tough the journey would be, we would one day make it to our ancestral home. That did not happen as the war dragged on for years

We were later told by the SPLA soldiers that Sudan had become a safe haven for members of Muslim brotherhood. Thing became so bad with the fall of Torit, Rumbek, Bor, Bentiu and Tonj to the enemy. There was report of Wau being besieged by SPLA and that was not enough. News of any liberated area would spread quickly and be received with rejoicing. The International media such as the BBC Focus on Africa report on Bor, Kapeota, Torit and surrounding outposts being taken by the enemy increased our hopelessness. There were losses of lives and properties with the fall of these towns.

For example, hundreds of thousand people were forced to flee to neighbouring countries as refugees. The civil war was framed as the Muslim north against Christians in the south. Civilians such as women and children were caught in the middle of Africa's longest civil war. With the war fought on the Christians' land, many villages were torched to ashes.

And when I heard sounds of machine guns, I would think of my scattered parents and brothers. Being the youngest in the family, I had expected that only death would separate me from my parents.

In Dinka society, the love accorded to first child is equal to the love given to the youngest child. With most women enslaved by the Janjaweed for horsemen, there were many concerns with the fall of many towns in the south in mid 1992. To draw enemy attention, Dr John Garang ordered the SPLA to attack Juba resulting in many cases of extra-judicial killings after the SPLA was repulsed.

The militia drawn from ordinary Arab nomads used the stereotypical idea that Africans deserved to be mistreated and oppressed. There were egregious serious crimes against the civil population committed by a combination of pro-Arabs militia around Juba and Kapoeta. People like Alaak from Kongor had his body brought to us. He was killed by a militant recruited from Torit. With his cousin, who later escaped, he was ambushed at eleven pm near small valleys between Natinga and New Cush. He was shot around the chest and killed instantly. The militant needed nothing from him other than his AK47.

Alaak, as we looked at his lifeless body, was said by his comrade to release a bullet which also blew up the skull of his murderer. The two were laid to rest on the same day. He revenged his blood and died as hero and his comrades mourned him. Attempts to locate the relatives of the murderer from the Dingdinga hills by the chief was in vain.

We had to cross long distances because our movement was fuelled by fear. Children who had left loved ones and family members such as in case of my parents and aunty Nyandeng Alaak would just continue to pray that nothing bad had happened to them. Aunty Nyandeng as an old widower was left at Piomahol. Her children all had died and she depended on our support. She was left defenceless.

In our company, were Konbin Bul, Manyuon Bul, Malek Arok, Ajak Atem, Aluol Ajak to mention a few. The bullets were raining beyond us and our team sometimes had to walk with groups of strangers. Because there was no available solution to my demands, the only available remedy to me was to rely on God's intervention. I was told in the village church those days that He looks after the oppressed and mistreated and was the father of deliverance. Having lost sight of my Mom and stepmother Nyanwut, my heart was shaken and my imagination of life as an orphan dominated most of my daily troubles. I would cry and stop talking.

Sometimes I had to be ridiculed by passing people who thought I gave up in life because of the associated hardship of endless wondering in the wilderness.

I would think of the whereabouts of my parents. This was emotionally suffering from my inner heart. We had to soldier on. By four pm, I trekked the several miles heading to unknown place of safety. We also had to make sure dogs and cows were the first removed while approaching Bor from the movement. They made noise and could make the whereabouts of people known by the enemy. For example, dogs have a tendency of sensing danger and could alert people to run or be followed by armed militants. One must come with possessions though hens were allowed. One of the people who came with a donkey was asked to smother its anus with oil so that when the donkey hee-hawed, no powerful sound would be produced. There was saying oil softens muscles around the anal area and no sound would be made at all as air escaped freely.

The light sleepers would tie ropes around the ankles of deep sleepers so that when any commotions arose, they would be quickly pulled to get up. Cold water was also useful. We were also advised that when one hears the sound of guns at night, the direction would also be confusing and one should not run away. One must first ascertain where the sound was coming from before determining the cause of action.

When the night ended successfully we would worry of what would happen the next day. And during the day, I would see flying vultures attracted by near dying children sitting on the branch of any available trees watching. Sometimes you found young children as five unable to move and needing to be carried. It was survival of the fittest as the most malnourished and sick as people fled the internal displacement camps with the coming of the Sudan Army Forces.

Children as one year could not understand why parents prepare for them leaves of grass instead of milk or fish soup. I had to witness even adults dying while calling out the names of the very cows left in the village. Hopelessness and depressed life marred most of our journeys. Sometimes I would wonder why the vultures followed every step we took. As birds of prey, the vultures had the intention of following the walking dead. With malnutrition, the cases of people appearing like skeletons attracted more predators.

The group had strong ones which were charged to look after the impotent. They would cook and distribute food first among the sick, women and children. There was less concern on interactions among the members of opposite sexes because everyone was worried about security not about making families.

At cattle camp as I knew those days, there was a right time for mating. Those times with plenty of food were unique. Encouraging the heart fainted and lies were also useful in most of the night journeys where distance expected to be far would be painted as the shorter distance. Those who had knowledge of the area would say anything about the views of the mountains appearing at the horizon. Those who became sick of diarrhoea could not get the best help. Other health complications were many and we had to sadly lose our little nephew, Kamich Abuoi Thon. Kamich died at the Ame displaced camp. I received the news while at Malek.

There was no border between life in the forest and in the displaced camp. Besides, there where enemies monitoring every moment. They would also rely on flying birds for there was assumption that where there were birds, there were human beings.

At my back was horizons which marked the end of no man's land borders and the beginning of endless dense forests. It was like everything about life is over but we had pressed on through the lean pathways. My feet and knees were throbbing but I kept on walking. I knew that surrender was to the government forces near Tali. And there was no more thinking of retreating. At a young and delicate age, I had stepped onto a bit of each corner of the former three regions of Bahr el Ghazal, Equatoria and Upper Nile. It all started in Aliet, Tali, Maridi and Narus. Without my father, I had enough in my mind of stories of the mistreatment of the orphaned during evening riddles and storytelling back in the day and I had to keep the impetus. For instance, the short stories would start from two sisters who migrated to Toch. As a result, they relied on begging for bones until they knew how to catch fish and swim.

We found many villages deserted. They were instead becoming desolated places because the enemy was entering near by towns. One retired pastor called Jacob would then call us to come together for prayers. It was our weapon. He would encourage us to forget about our nice houses and cow milk.

Not all of us listened to good advice when soldiering with realities of harsh environment. Some children would not understand why they must walk at nights and spend more days without food. They would not even understand why the sound of guns became endless.

Some would cry until they were given what we used to give them in the village. We would organise short prayers in the evening under the trees. Three things were important: his protection, guidance and health. Sometime the prayers would be very long but touching every moment of our problems. One day Jacob had to decide that prayers would be circulated among all members. Then came the turn for five year old girl to pray. She began her prayers like this:" God of the suffering people, could you return us back to our homes." Then she burst into tears. She had separated from her parents. Jacob had to finish the rest of prayers and made a quicker conclusion of Amen. We had to comfort the little one.

Her prayers were powerful requests to a supernatural being. She intended to see intervention from above. Having children as young as five in the group was painful and disturbing. That little girl did not know whether her parents were in the village or in a mass of moving terrified civilians fleeing the aerial bombardment. She did not even know that the government was occupying people's land and terrorising everyone including mass killing. She was referred to as an angel and many people liked her before her parents found and identified her after three days with us.

Inside the forest we made as our home, there were people who became homesick. Children who got lost and were cared for by well-wishers began to call for their dead parents by names every night. People who got thousands of cattle heads later taken by looters would count the number of bulls while also narrating terror in their dreams.

Those who had never left the villages since time immemorial would blame themselves for accepting to escape from fighting other than to subscribe to call of death. Without nothing left, the world of Dinka who kept hundred heads of cattle was very dark and hopeless.

There were people who owned hundreds of cattle who reported of looted cattle. We had received our report while on the move and we were expecting nothing. Our concern was about safety of my family members starting from my mom to my brothers.

So, in my four am private prayers, I would pray for all my family members and myself. It was the whole house wandering the wilderness. For instance, Ayual and Mayen had to flee to Ethiopia having been dislodged following the fall of Regime of Mengistu Haile Mariam in May 1991. They were the lost boys of Sudan known for walking thousands of miles to the destinations of their leadership choices. They were unaccompanied minors offered by parents to join the movement when they grew up. Some included those who assisted me when I was drowning in a small pool in 1985 such as Deng Achol who perished in Ethiopia.

Trained as child soldiers, most of them had skills from building small huts for themselves to carrying guns bigger than them before they left Ethiopia.

On the way, we arrived at the Nile riverbank and passed through the area which was a resting place for people who came from fishing expeditions. Laying on the grounds were empty boxes used by the government army to carry their ammunitions. We took some to use as cookers and containers for group eating. This time I was tired for I had spent days without proper sleep. So, I fell asleep while walking. I had at my back-water container. Its capacity was three litres. It was better to carry a small quantity of food than to leave the water behind. We had in mind that there would be vast areas with no water sources. In those days, I was told by a friend that if one wants to go hunting, the first thing to do would be to get enough water.

In our time of displacement, there was no leisure but soldiering on through the alien jungles amidst fear of unknown eventualities. To completely leave the enemy behind, we had to move quickly. Attempts to sit or rest was not possible because of the conscious guilt of seeing many wreckages of military tanks destroyed during gun battles.

It was one of the darkest nights. The combination of darkness and exhaustion hindered my visibility and I could not see the abandoned truck during the dirty road between Jemeza and Mangalla. Accidently, I almost lost my teeth. It was like someone had hit me with electric rod. I stood still for about two minutes to regain my strength before I soldiered on. I increased my strides to catch up with the second group. I was almost last and was already behind the group.

While walking along the Nile bank, we could smell the breeze emanating from the river and felt like there was both relief and life. In those days, many people would seek shelter in the bush as a cover from being spotted by unknown adversaries.

People like an old woman Achol who was disabled by a snake bite were trailing behind the group. Due to dawn raids by our attackers we had already left the Twiland behind. I could only see smoke from grass houses lit by the enemy to increase the fear in us. In a narrow escape, each one fled in the direction the force of nature dictated. I met some people hiding in the bush that narrated that they had to struggle with fat bedbugs. As the rest headed to the Ame displaced camp, the number of chigoe flea or jiggers infested feet were unimaginable. I had three big ones of hundreds of thousands of eggs.

As the increased dirt covered few clothes, there was also an issue of where to get soap. Instead others would resort to the use of only water and sunbathing to kill the insects. This was done by spreading the cloth on sand with waist facing the direction of the sun. When it became hot, the lice would temporarily disappear. The remaining ones had to be handpicked.

It was tough and without a comb my hair grew in the direction of its own choosing. The hatched larvae dotted the dirty cloth and the dead ones would lie white attached in between the seams of trouser joints. There were people good at blowing up the bloody ones by crushing them with fingers of two thumbs. Then patches of ringworms like a ball began to appear among our group. Instead of cow dung ash to whiten our teeth, we resorted to any available soft leave of tree to remove maize particles. In the process, some boys developed tooth decay and could not properly chew maize.

Without electricity to guide our feet, we would walk and cross through as long as the moon shone. Things were different when walking without the moon and with the help of moonlight. For instance, one could not tell who was coming up from behind to snatch. I used to hear from my brother Mayen and Deng Wal after we met that there were cases in Ethiopia where some of lost boys got snatched by unknown monsters while walking the paths from Ethiopia to Narus. Deep in the jungle was fear of many human eaters though they were scared by sounds of the guns. There was also belief that a rooster crowing at the wrong time was an imitation of bad omens.

Laying by the roadside before we approached was a set of fresh bones of a woman. We looked on as we proceeded.

The brutality of Mongalla could be tracked to the period of Lado Encave of King Leopold II of Belgium. His reign of terror was not confined to Congo but was also extended as far as Greater Bahr El Ghazal. His conquest for minerals and exploitation of Africa marred the period between 1899 and 1904.

By the time we passed through the area, there was fighting behind us between the SPLA and government militia. Then we rested for one hour. By seven am we had already headed southward of Terekeke. It was sunny weather. Sometimes we had looked for shelter when it became too hot. Talking was possible if we were sure no enemy was following us. The best time for us to move was at night. It was because it used to be cold at night and saved our water. Second, the enemy would only identify our feet in the morning after we had gone. We were so hungry and I left the people sitting at the big tree. I took someone's spear with me. I jumped into a small swampy area which was between the Aliab and Moro vast border. And luckily I killed ten kilograms of mud fish from the river. This added to some of the food provided to us by the chiefs of the area.

We spent a night before arriving at a certain SPLA outpost between Yei and Mundri. By eight am, we had crossed the broken bridge and made it to Mundri via the airstrip southwest of Lui.

In this company, we had a full escort provided by Rock who was the officer in charge of Tali. We spent about four days in Mundri before CDR Garang Mabil provided us with his hardtop land cruiser to take us to Maridi. We passed through Kotibe which later became a camp of internal displaced persons. We rest for an hour and made it to Mambe then Maridi.

We arrived at a SPLA manned roadblock at ten pm. We then intercepted a security code which used to begiven to the town that night. On our left-hand side and first upon entering the town was a Mosque. Our arrival was on 10 April 1992. We passed via a small market and proceeded to what once was a Sudan Police station. We were taken to the house of 1st Lt. Deng Ajang Awai, where we were looked after by his wife Nyanwut Koryom and Clementina Kombo.

We spent a few weeks and were then offered another house. In June, I joined a school run by Catholics. It was near a Catholic hospital and the Catholic Church. I would cross a small bridge east of the hospital and walk for about fifteen minutes. Our house was an old building left by Arabs who got ousted from the town. Bullet holes on the roof were still visible. The market was three minutes walking distance from the southwest direction. We had the privilege of playing handmade ball at big empty field. The field was used for the SPLA parade and celebration of 16 May 1983.

At school, I was introduced to the Arabic language. After we had our parade and sung war songs, we would welcome Comrade Atem Biar with a salutation. Then he would start our orientation on the political ideology of the movement and its objectives. He would allow us to cite the SPLA/M allegiance in Arabic which ends with "SPLA shall win". Having made a proper introduction to us, Atem was redeployed and I found him at Chukudum in Eastern Equatoria in 1996.

In May 1992, the SPLA comprising of Commandos and other specialised forces arrived at Maridi. They came with officers who had fought and defeated Sudan Army forces in most of the conflicts in Western Equatoria.

They had with them few tanks and machine guns. They convened at our home. Among them were Commander James Hoth Mai, my maternal uncle Abuoi Arok Deng, and Garang Madiing Agok among others.

To greet them, I saluted before we had warm handshakes. As a potential child soldier, I had to act like someone who had had prior proper military training. The army's salutation was precise and excellent. In return, I got a nod of acceptance from more of the sitting officers.

Before they left for Juba in a week's time, my uncle ordered some of his bodyguards to remain behind to help in the cultivation of maize and groundnuts. Kuir Wieu and Kuir Kiir and Dut Aleu were left behind. He went with Akech Mel, a muscle machine gun operator.

Malish who was told to remain behind was not happy. There was a prophecy of the local prophet Bith in the company of Garang Madiing Agok who foretold what would happen in Juba. Everyone was inspired by his previous prophesies. The morale was remarkably high from some of the soldiers I met. They considered those prophesies came true with the defeat of Sudan Army force at Bo Bridge.

At four am on 16 May as usual there was a shot. It was soldiers' celebrating nine years of 16 May 1983. I woke up shaking with thought that it was an enemy attack. This was the day revolution took off and it deserved commemoration with gunfire. The Sudan Army Forces as uncle told me was not waiting for them because there was information from Dr John Garang De Mabior of Sudan Army Forces intention to retake Torit and this plan had to be foiled. They were ordered to leave Maridi for Juba with immediate effect.

Before the end of May, I could not see many soldiers. By June, the SPLA was engaging outposts around Juba before they assaulted the town. They occupied the military headquarter, Giadia, penetrated deeply to Kololo, and Konyokonyo and besieged Juba from all the directions including Bridge on the Nile.

They blew up arms and ammunition storages. Back in Maridi every one of us was sitting near the old radio listening to the BBC on the success of the attack. Our prayers were that Juba must fall. The SPLA inflicted heavy losses on the enemy with casualties on their first assault to the town on 6 June 1992 despite the lack of coordination coupled with inadequate machine guns and tanks. A few days later, we were told the SPLA was repulsed with casualties, although Juba continued to be besieged. People like uncle were pushed back to Korwilliang near current Jebel Suka and they had to shell Juba from a distance. There was continuous fighting at around Lafon. Several months later my uncle arrived in Maridi on a mission. We were excited as we received him. No one discussed the issue to do with what happened during the attack until 2006 when I decided to write a book.

He was among the commanding officers who made a successful incursion to army headquarters before enemy fire pushed them back. By October 1993, we left Maridi. Then we passed through Bahr el Laam and Fakethika. It was in this area where my cousin, Dabek Mayen Ajak, who was an SPLA soldier was killed before their forces liberated the area. He remains an unrewarded hero.

We proceeded to the north east of Maridi and spent a night between a narrow muddy paths branching to to former Zaire. By nine am, we had already arrived at Bazia.

This outpost at the border had rocking Congolese music and people there were so happy as if there was no war.

We had to shake our legs with the Congolese music and that was after we had sold important items (plastic sheets) in our possession. Now, we had to infiltrate the enemy line of defence since Sudan Army and its local militants used to lay ambushes between Morobo and outskirts of Yei. We had locals showing us the directions and positions of the enemy. It helped us to pass.

Then we heard artillery shelling and we had to cut through the forest. By four pm, we found SPLA reconnaissance who aided us passing near the SPLA trenches. At about five pm, we arrived at the SPLA base. We found 2nd Lt. Alaak Thon Kamich, who had injuries during the Maridi offensive dug in. He provided us a place to sleep and advised that we run to the bunkers if there was artillery fire. In our entourage was Kuir Kiir Reng.

Kuir had in his possession an AK47 and a hand grenade. Early in the morning, we left for Nyori. We had to narrowly escape bushfire and exploding explosives left on charred battlefields. Nyori was in ruins after aerial bombardment. We slept for two days then proceeded to Mangalatoria having crossed Yei River while it was still dry. Yei River is one of the fastest moving rivers in the area and we were told crossing it when it has water is done with ropes.

After two months, I joined the under tree school at Mangalatoria and spent the entire December up until April 1994 there. On 23 April 1994, we left for Kajo Keji hundreds of miles north of Uganda. By noon, we heard the sounds of warplanes before officers in charge of air defence rang the warning bell. I was sitting under the mango tree at the compound of the late captain Akech Aguer Bul.

We entered the trenches. Scared of fire from the SPLA artillery unit on the top of hill, the plane did not bomb Kajo Keji. We spent a week and then crossed the River Nile with a small boat. Then we passed through a mountainous region which stretched from the river all the way to Mount Gordon near Nimule. We had to nurse those who could not make it at the peak because of exhaustion.

By ten am, we arrived at Pageri and headed to the SPLA base which was manned by a majority of SPLA soldiers of Nubian origin. I was shown where our cousin comrade Duot Alaak Kuot was said to have been killed. Duot was one of leaders of Red Army who was ambushed and eliminated by unknown assailants.

While we were making a decision on the next course of action, there were rumours of serious fighting in Lafon. We were later joined by Manyok Ajak Chieng who left his forces with permission. He had just lost his wife. There were reports that the enemy would intensify attacks on all the SPLA controlled outposts around Juba to flush it out from all the liberated areas.

They were also closing in from various directions and Kajo keji. At Pajeri, I befriended many fighters mainly from the Nubia Mountains. Some of them assisted me on how to disable AK47s and took me around ammunition storages. A friend called Ahmed would open boxes which contained howitzer gun (M777) missiles and how they are loaded. In the evening, I would pass through the trenches to fetch water from the tap which was about 200 metres from our place. And before I moved, I would make sure I knew the security code for the night. It was common practice to tighten security at night because no one knew when the enemy might sneak in and take over the town.

The ruins were an empty hut built to host the SPLA/M Conference of 1994. It was dug deep to provide protection to participants who were drawn from all the regions of Sudan. I went inside by mistake. In a few minutes, I saw about five men armed sent to arrest me. They ordered that I should come with them to their commander who sat under the tree waiting for group lunch. Upon arrival, I found six officers playing cards.

I was asked to stand in the sun, interrogated and forgiven having known that I did not intend to do anything wrong. While taking their jobs as matter of life and death, some of the intelligence agents reporting on anyone perceived as supporting the enemy could attract a death sentence. I was cleared as a genuine supporter of the movement.

I was later told the place of conference was changed to an undisclosed location at Mountainous zone around Lotuke (Moyothukul). The conference was to discuss the issues of the liberation and adherence to the principles of army struggle. Commanders like late Lual Diing Wol and Yousif *Kuwa* Mekki were instrumental in the deliberations of the objectives of the struggle.

We later heard that SPLA/M Dr. John Garang was reconfirmed as the leader and deputized by Commander Salva Kiir Mayardit. The criteria used in assessing individuals to occupy the position included unquestionable loyalty, consistence, and an oath of allegiance to even die for people's freedom.

By May, I left Pageri with Garang Ajak for a more secure place. Later on, we heard Pageri was deserted. It was unfortunate I got separated from Ahmed, the gunner.

In a few months the enemy retook part of Aswa few miles away and Pageri. The sounds of artillery and aerial bombardments were terrifying. The commanders would make sure that when enemy approached, its strengths were assessed. If it was deemed as costly in both humans and resources, then a tactical withdrawal would be ordered. Pageri was evacuated southward and forces were concentrated to defend Aswa.

Then I crossed for the second time to Mangalatoria. I stayed there until June 1994. The area was dotted with unexploded ordinances and an enormous number of destroyed, charred Sudan Army vehicles defeated in 1989. Some of these vehicles laid overturned. I was told there was vicious battle at Khor one hour south of Mangalatoria. The officers there pointed to where SPLA laid ambush against the forces under the command of Isaiah Paul. Paul was in Sudan Army Forces though a Southerner. He was on the side of Sudan's Army because of his knowledge of the trenches.

There were many dried bones lying on the surface as well. Then someone among the officers pointed to the horizon and said that that was where my cousin Bol Akoy Bior was killed on the side of SPLA. According to the witnesses' accounts, he was killed by a wounded Arab soldier during search and clean up operations.

On 25 June 1994, someone came to me while I was sitting in a shanty market. He was acting under the instruction of Biar Manyang Jok who was a lieutenant. Biar was the machine gunner operator of Koryom. He is skilled fighter who had experienced fighting in many battlefields.

Figure 8 at Juba Grand Hotel 2010 with Biar Manyang Jok, Hon Atem G D Kuek and I

He is best friend to my maternal uncle Abuoi. He was trained in Ethiopia and then assigned to commando. The person advised that I should get ready by the morning of 26 June 1994, because there would be traveling to Eastern Equatoria which would first start with crossing to Nimule.

As agreed, I woke up the next day about six am. We left Mangalatario and by twelve pm when we were about to approach Kajo Keji we received the news of the fallen town.

Deafening exchanged of artilleries between the two forces now made it even more worrisome. We could hear the sounds from about three kilometres away.

We had to divert to west of Kajo Keji through the deserted villages of Jolmo, Pomju, Kaju at the border of Sudan-Uganda. We met SPLA soldiers dislodged by the enemy at the water point. We greeted each other. We took water and proceeded. By eight pm, we erected our tents in the midst of the forest. There were no people in most of the villages. We cooked maize and ate it.

By four am, we left following the bank of the White Nile through Ambo. We could see the landscapes of Aswa on the side before we crossed to Nimule National Park. We were six in number. Then we took a wooden boat and crossed the Nile to Nimule. It was already about two pm. The Nimule waterfall which we just found had drowned a giant hippopotamus.

The speedy water was rolling it many times which hindered our efforts to retrieve the carcass for our diet. Konbin, Manyuon and I spent about two to three months in Nimule. Nimule was among the few town under the control of SPLA and bordering South Sudan and Uganda.

Konbin, Manyuon and I stayed there under the care of Biar's family. There was starvation and people had to survive on cassava and wild fruits. Sometimes, I would visit my cousin the late Makuei Chol Dengfour. He was trained in Cuba and sent back. Makuei would make sure we eat if he had anything in his disposal. Because of starvation hungry people would have no option but to eat weeds. Some were poisonous and some lost their lives.

Then we made the decision to cross to Labone though there was a restriction of movement since the SPLA was in operations around Aswa, Palataka, Pajok, Pogee, Lafon and Ame. In one afternoon, we received our departure orders from Commander Abdel Aziz Hileu. Step by step, we left Nimule and got to the Mangali displaced camp. Hileu was a courageous and brave officer who had a vision of liberation at his fingertips. Before we reached Mangali, we slept at his camp inside the forest. He might have been there purposely to counter the enemy flank. We were served with hippopotamus meat by his soldiers. Then I recognised my cousin Abuoi Deng Mabior with about ten young men of Jesh Amer (Red Army). They were being properly beaten and rolled in the mud.

I moved closer to the officer in charge and asked what they have done to receive such forms of punishment. His replied was simple; they went to the forest, took someone's goat and ate it. While I was about to move closer to them, they were already pushed into fox holes. The next day we left six miles from the Mongali internal displaced camp.

It wasn't until 1999, that I came across Abuoi Deng again in Narus. While waiting our order of hump, I asked him what he had done that he was beaten a decade ago for. He laughed before he said they were hungry and waiting for LRA (the Lord's Resistance Army) and they could not allow themselves to die of hunger. So, they snatched one of Abdel Aziz's goats. My response was that you had a gut to touch what belongs to the commander.

In front of us was the Aswa River. It was a river full of water. The enemy had besieged Nimule from the north (Mt Gordon) and Aswa. It was only a matter of time before anything happened to the two towns. However, the SPLA soldiers under the able commanders of Oyai Deng Ajak, Atem Aguang Atem, Deng Madot, Ajak Yien, and Awet Ajing among others had a life or death resistance against the enemy. Landmines including hooks were planted inside and outside of Aswa River and its surroundings were defended in defiance of all sorts of enemy bombardments. In the evening we would hear aerial bombardment.

At Mangali and to cross Aswa River, we had to look for a wooden boat. Successfully we sneaked out of the river far from the sight of the enemy. But there was another threat ahead of us. That was the Lord's Resistance Army of Joseph Kony. The area we were crossing was its infested jungle. The savannah grass was three metres higher. We could not see anyone approaching.

Before we arrived at Pogee, we had to cross small tributaries with fast flowing water but we had to sleep in the forest. On the road, we found footprints of the Lord's Resistance Army. But we did not bother that much because some of our team members had guns with them.

We bypassed Palataka from its southern direction and headed southeast. Tried and starving, we took maize from a certain farm. The owner might have run to Uganda because of impending attacks from the Sudan Army and the notoriety of LRA. The villages around Pajok and Pogee had experienced constant raids from the LRA a few months earlier.

We roasted the maize and asked for forgiveness from the owner because we had heard stories. A soldier had eaten a rooster of certain chief with magical power. He took it without permission only for the rooster to make it's normal morning crowing. That news was disseminated across to every soldier or any stranger, so they would be careful with taking other people's items without permission. Having asked for forgiveness from the owner and God, we slept around the fire. It was a cold and cloudy night.

Pogee was not a big town. It did not have big buildings and roads but it was a strategic place near the border with Uganda. We had a rest and the next day, we took off for Labone. We arrived at eight pm the following day.

Labone was once used as a base for the Anya-Anya One in the mid

60s. Southern fighters of Garang Juach and the rest of his colleagues used it as launching pad against the Arabs. It is located deeply inside the valley and covered by mountains. Around its valleys are thick bamboo trees. Inside the bamboo just four hours before Labone, we found one Lord Resistance fighter. We suspected him as a notorious monster but he lied to us that he was a civilian. He became sick and was resting before heading to Uganda for proper treatment. I became suspicious but we had to mind our own business. At Labone, we built small grass huts made of bamboos. Then I started a bamboo chair making and handmade seamed plastic bags business. I stayed there until the Lord Resistance Army attacked the area from the north. People like Duot Gieu, Ater Deng Ater and Dhieu Akech Adier among other SPLA soldiers were sent for reinforcements.

The notorious militant had ransacked part of Labone killing people and looting anything they could find. Dhieu and Ater were killed via an ambush. The news reached us that Duot was wounded on his hand. The passing on of the very men who ate together with us just recently was painful and unbearable. We mourned them not only as relatives but those who had fight in many battlefields as Koryom. The civil war was had gone for eleven years. The LRA used to carry raids under the Sudanese government's instructions. They were based at Owinykibul and at the Uganda border. Sudan was engaging in a proxy war with Uganda which they accused of providing support to th SPLM/A.

We had to accept the loss. In Labone, I also met my nephew Atem Diing Atem (Achol). He was working as an SPLA soldier with medical corps. One Sunday, I was taken by soldiers with the intention to be conscripted to their ranks and files. Then Atem popped up from nowhere and I was rescued. By the end of October, we left Labone entering Uganda from Boro and Maduwufe. It was the first time I bought a nice looking pair of plastic sandals. Most of the distant crossing and journeys were on barefoot or with use of old dirty pairs made from old lorry tyres. Because I had little money from that bamboo making business, I bought soap and few clothes. The team had increased to include Arok Ayiik and his brother Ajak, Thuch Awer and Gai Ayiik, Garang Kiir and few elderlies.

To get through because there were restrictions, Mach Guguei had to intervene based on the request from Uncle Abuoi Arok. Having already crossed via Uganda, we spent two days between Maduwufe and the jungle before arrived at Tsertanya. One of our team members was sick and we had to wait for them for one day at Tsertanya.

Having had enough rest, we took off and slept in the forest. Here, we had over hundred soldiers meant to open routes to New Site the new headquarters of the SPLA/M Leader Dr. John Garang De Mabior. Here there was no attacking of the enemy positions but to cross. We had wounded and old veterans in our company.

After all day trekking, we passed through Acholiland before we arrived at Ituko. Then Tule, Lomothing, Amtaini and we spent a day at Kilkil. At dawn, we moved, and our first batch arrived at Lotiki. People were organized into groups to ensure spacing and reinforcement in case of an attack.

In most of the travelling, I rarely lagged back because being in the lead meant I got to take rest before those left behind for hours joined our group and moved again. Then, I bumped into Juach Kuot Deng. It was seven years since we were separated in Aliet. Juach was taken to Ethiopia, trained and posted at Lotiki. He was only awaiting orders with his Jesh Amer colleagues to attack Kapeota. Here, the command of the movement was moving troops around Torit and Kapeota, Eastern Sudan and the Nubian Mountains. The soldiers who came with us were mainly from Wuduk and under the command of Malik Agar. We were told they would be taken by cargo to Blue Nile to engage the enemy from the north. Being exhausted and hungry, Juach rushed to his store. Quickly he rushed to me about three kilograms of ground nuts. I gnawed some before I ran fire around them.

By evening, he brought food made of millet. Then we rest for few days. Lotiki is surrounded by ranges. It was surrounded with heavy machine guns and every soldier was on high alert.

Ahead of us was Lotukei and a view of Lotome/Nyala. While on the move, our clothes were covered by red dust, a sign of quarry activities on the top of the mountains. It was a fertile land with many species of trees and grass. Northeast of Lotukei beside the ranges was the khat planation.

The area was cold and acting under the time pressure as soldiers were under instructions to make it to Masawa (New Cush), we proceeded directly there.

In the morning, we had already made it safely. We spent a few days there. New Cush was SPLA's safe sanctuary far away from bombardments of Omar El Bashir. I and Mawut Bol Awuol went to Natinga but that was after I was taken by soldiers to do grass cutting around a big grass thatched twelve by fifteen metre house built to host the women's rights conference. Then I was detained at the roadblock before my luck came with the passing of Commander Rebecca Nyandeng De Mabior, the wife of the rebel leader. We were released immediately upon hearing her approaching.

With our belongings on our heads, we made it to Natinga where I spent three days at the grass thatched house of Garang Akech Deng Lou who was away on a SPLA mission. It is small outpost surrounded by tiny hills. In summer, the Toposa of Nyangathom would bring their cattle to Natinga because of its water reservoirs.

There was no hair in the middle of my head. This was because since I left Maridi, I had been carrying my belongings either on my head or my back. Some people would do both carrying AK47, ammunitions, bags, food items and a child.

Then I met maternal Uncle Manyok Ajak Chieng who was responsible for those boys displaced from Palataka and in school under Face Foundation. At Natinga, we would go to the forest to cut trees and grass to make small huts. We built huts and I stayed there until the morning of 10 November 1994. While at Natinga, we constantly received the news of Kapeota failed offensives. Many of the Red Army (Jesh Amer) fighters were unfortunately killed. The SPLA lost its brave commanders Luol, Majok Mach Aluong and Anyar Apiu among others. We were all in a bad mood that day.

At Natinga, we had two goats and Dingdinga handmade tobacco. There was movement of people to Kenya and Sudan. So, I decided to join my brothers, Mayen and Ayual who were in Kenya. I gave the tobacco to the lorry driver as fare. He accepted the offer and allowed me in with Nyajok Chol Mabil. The two goats had to remain with friends.

It was 28 November 1991 and I last met Nyajok's husband Deng Garang Dau at the outskirts of Bor after Nasir forces were repulsed from the town. Sadly, Deng was later killed by the same faction while retaking the town of Panyagor from a combination of Riek Machar forces. I was under Nyajok's care.

She is a caring woman. While at Ladapal, she cooked for me. By four pm, we arrived at Lokichiogo, a Kenya town, at the border with then Sudan. That was after we had passed through New Site, Nakodo, Napadal, and Keybase. We arrived in Kenya for the first time on eleven November 1994 where we found the Kenyan Army patrolling the border. They stopped the lorry and thereafter conducted a security search on each one of us.

Then I told them in my poor English that we had been displaced by fighting and wanted to take refuge in Kenya. Then we were taken to a compound run by the SPLA humanitarian wing.

The officer in charge took us to the UNHCR compound where we had to be cared for by the Kenyan case manager, Kamau. Kamau though he helped thousands of Sudanese refugees mainly from Southern Sudan to settle in Kenya, was known for being ruthless and rude. At nine pm, he took our details and conducted some interviews to ensure we were not those who abuse the process by registering twice or thrice.

The process took an hour before we were squeezed into a blue lorry. By midnight we arrived at the Kakuma refugee camp (fenced enclosed accommodation), which could take over a hundred families. The fence provided protection and it felt secure and sleepy so I cleared the bush and slept on the floor. I did not have anything in my possession. The camp was established to host Sudanese unaccompanied minors commonly known as lost boys. People like my brothers were unaccompanied minors and soldiers who were displaced with hundreds of thousands of their colleagues from Ethiopia in the 1990s. Kakuma comprised of four parts namely (Kakuma I-IV) managed by the UNHCR with the Kenya Department of Refugee Affairs.

By eight am, I sat by a road heading to Kakuma town south east of the protection area where I spent a night. Luckily enough, I recognized Dengthi Garang Dau. He did not know that his brother, wife and I had made it to Kenya. All thousands of miles were crossed without any means of communication.

We had just appeared in a very dirty and dusty marred concentrated settlement like ghosts. Kakuma became home to almost half million refugees from the horn of Africa and Congo.

Along the landscape were tiny houses made of blue reinforced 4x5m plastic. The plastics were provided together with long timbers, with nails and ropes.

Also provided to refugees were crushed dried leaves of the aloe vera plant. We hugged. Then I asked Deng to tell me the conditions of my brothers. He said that Ayual had left the camp in 1993 with over eight hundred lost boys to be retrained somewhere around Lotuki. After his training, he was taken to around Mogri and then around Bor through the desert.

They fought once with one of his senior commanders, who rebelled against Dr. John Garang in 1992 at his base at Pageri before he reconciled with Dr. Garang four years later.

Racing with time and eager to see, Nyanjok, who was in a women's section of the compound, Deng had to excuse himself. He rushed to brother Mayen who was in class. He was at Aweil primary school before he skipped one class for Rumbek. Located inside our zone was the Shambe Primary school named after a small port located in the Sudd north west of Bor.

It was stunning for Mayen. He could not believe I was alive. He left his favourite mathematics session with teacher Wac Duot Wach. He hired a bicycle which was the first class mean of transport those days.

By nine am, he arrived in the compound. He found the UNHCR assigned representatives with their white lorry. Our names were taken and each was asked which group was their preference of one choice.

There were over fifty groups, with each comprising of thousands of persons. I was taken to 50B. The group was divided into A/B, like Pakou, it was predominately Ayual community people with few from other communities such as Palek. Mayen had with him baked cakes.

He gave the cakes to me. I ate it while inside the lorry. Nyanjok was called from the women's section and we were taken to group 50B. The people were in groups though more sectional than reflecting the composition of the camps, which was home to refugees from Sudan, Ethiopia, Somalia, the Democratic Republic of Congo, Rwanda and Burundi. The place is located at extreme far end of Northern Kenya and hot and dusty. It is semi-arid with varying temperatures of 39-40c.

It is part of the Turkan district with its county headquarter, Lodwar. The Turkana people are Nilotic like distant cousins to the Dinka, Nuer, Lou, and Nandi among other various groups of Nilotic speaking people. They rear a few cattle, donkeys, camel, goats and sheep. Historians had reached the consensus that Turkana people entered Lake Turkana from the north. They have acquired fighting skills following internal conflicts during the Ateker confederation disintegration hundreds of years before.

The Ateker confederation was a combination of Jie, Nyangatom, Toposa of Sudan, Karamojong and Teso. Their experiences in fighting exposed them to exploitation by the white settlers during WWI and WWII.

At group 50, everyone was waiting for me. I immediately met relatives and friends we had separated from going back to 1984. Some were SPLA soldiers, who retreated from fightingto the camp.

People like Alaak Deng-matiok, Lueth Deng Baak, Alaak Lueth, Mama Akuot Gieu and Achol Goch were ready to do the welcoming to the group. I was registered as a refugee. I began to get food rations though

Figure 9 Alaak Lueth Deng Baak from his family album 2009

a supply for survival. I was given secondhand clothes and a blanket. I continued with my worn-out pair of sandals.

I was provided with poles to make myself a house. We had to cover them with mud and plastered sand. Three weeks later, the schools were closed. I missed class that year until 1995. Near our group were the Wau, Cush, Imatong and Sobat Primary schools. They were full to capacity and hence making indistinguishable noise from bush schools in the rebel areas. So, we had to complete our first semester under the trees.

People like Kuot Jok, Garang Chol Jr who is brother to singer Garang Majumaar, and Dengawet, began to show their ability to lead the class. Confused between Arabic and Kiswahili transition, I had to put extra efforts to have my name heard in the top ten. I was doing extremely poor in Kiswahili beyond any repair. I was alien to Kenya having joined those who had been there for a few years earlier.

First I was assessed and enrolled in class three at Wau Primary school. With my age almost over fourteen, I was not happy. So, I left for Ngundeng primary. I jumped to class four. The school was thirty minutes walking distance at the beginning of zone five.

There was another problem. I was considered as an intruder. After one month, I was joined by my cousin Alaak Lueth who was also jumping from class four to five.

About two hundred metres from the school was a market where all sorts of items were sold. The drunkard would fight, and staggering thugs would disturb our lessons with stone throwing.

I did not have any certificate to prove that I got promoted to class four. And when pupils were called using the registry, I would walk out from the class pretending to go to the latrine. In that roll calling avoidance, I spent the second semester and I did well in examinations. I was provided with a certificate.

I became a full member. Having achieved what I had intended, I left

Ngundeng and I moved to Imatong Primary school. I met head teacher Kuer Dau Ngewei a kind-hearted person who used to assist many naughty students to achieve their desired dreams.

He was not type of a teacher who flogged pupils but mentored them to be higher achievers. In most of the schools, a lack of discipline was common. For instance, some students would wear trousers instead of shorts against the school regulations. Some strong boys acting under the influence of adolescence were fighting teachers. But what most students did not know was the fact that some of their teachers were SPLA officers assigned by their leader to build the next generation of future leaders.

To reinforce discipline, there was hard rubber warped out of truck tyre to flog students found to be non-compliant. It was in possession of teacher Malaak and was dubbed as black uncle. Small or big would run to class upon hearing that teacher so or so had removed black uncle from the office.

With the help of my brother Mayen, Kuer allowed me to join his school. The school was five minutes walking distance from my place. I completed class and I was promoted to school five. The life in the camp was hard and difficult.

People would spend days without food and there was relative group to group violence instigated by idle people. This made the camp no great difference that being with soldiers. Food distribution was 2100 calories per head and there were months when the camp could go without food.

I completed class five, but I left in December for Natinga. I had twenty Kenya shillings with me. The minibus used to charge ten shillings from Kakuma to Lokichiogo.

I was left with only ten for the entire journey to Narus. Upon arriving at a second police station before Lokichiogo, I was asked to come out. The police shrieked at us as Wa-sudo literately Sudanese with intention to force us out of the bus. I came out with Chol Juach Young. Then we were taken to a dirty confined small office, interrogated and threatened with guns.

All sort of intimidations from entering Kenya illegally, being soldiers, criminals to name them were thrown at us. As a result, the two policemen asked that we bribed them. The bus was still there waiting for us with driver chewing khat known locally as Miraa to keep him going. Miraa contains the stimulant factors of alkaloid cathinone and causes happiness and loss of appetite.

The driver pretended to know nothing. Between Chol and I was another Sudanese giant guy with long teeth. One of his teeth was curved upward like that of warthog. To ensure we pay out quickly under duress, the police grabbed the stick and beat the guy. He was like, who told you, who told you to laugh with such teeth. As a result, the man struggled to cover his mouth. This did not help him and every one of us. Finally, he said it was prohibited in Kenya to have such teeth. In a refugee camps back in Kakuma, people had no access to dental services as Turkana district was one of the poorest in Kenya. But for the police, it was his luck that these poor refugees had to feed his belly. The drama lasted for fifteen minutes. Then they dug their hands into my pockets and found only ten (10) and they took it.

By the time they counted their loots, they had hundred Kenya shillings taken from us.

We were released but because my money was minimal, the police gave me a kick. Then he released us. We rushed into the car. By about three pm, we arrived Lokichogo. We slept at Link ran by the SPLM before I took a lorry used by cattle traders to Narus.

I spent the whole of 1996 between Narus, Natinga, New Cush and Chukudum. In Natinga, I became sick. So one day I decided to trek to Chukudum to seek treatment.

Without a departure order, I was briefly arrested at the Natinga roadblock to New Cush. By noon, a truck with Norwegian People's Aid labelled arrived. Having been released and under the tree, the driver agreed to take me. We arrived at New Cush. The following day, the driver refused to take me to Chukudum unless I paid with a rooster.

He left and with no choice, I went to the nearby SPLA post. I met SPLA soldiers and group of mafia like Agoro traders with hundreds of heads of cattle for sell in Uganda. I was provided with an AK47, we took off. We would sing songs which send away road attackers. And in the evening, we arrived at Chukudum. The Agoro was established from former soldiers and armed to teeth. As soldiers they would herd their cattle in columns and were systematically organised that no number of raiders could attempt to infiltrate them. They had commanders and could pass through LRA infested areas cutting between Kilkil and Chukudum on the way to Sudan-Uganda border. We separated as I had to get to Chukudum with four people going for treatment.

After an hour's walking distance, I found an army truck. We were told it had spent two days in the middle of the road. At the rear were two traders bitterly arguing because one was accusing the other of stealing ten litres of wine known as KK. The driver intervened and I could hear him saying that no one would drink such amount without becoming drunk. Even before we left the truck, the wine had worked on the accuser's blood stream and the thief was exposed.

On my first day at the outskirt of Chukudum, I found many bones. The bones belonged to combatants killed during between the SPLA and militants in the 1990s. The bodies were left in the open. Under the mango trees next to the roadblock south of Chukudum Town were skulls and bones inside the trenches.

I stayed there until June 1996. I sold salts that I had brought with me. Then I bought a full sack of tobacco. Then I heard that there was Hino truck full of ammunitions and carrying 200 litre drums of fuel which was about to go to Natinga. The truck was under the command of Mohamed from the Nubian Mountain. Mohamed was my friend. He had only three fingers having lost two to shrapnel while defeating Chukudum.

By ten am, we had left Chukudum. We passed Kilkil before the Hino overturned between Lotuke and New Kush. All the people inside survived and only one elderly man had a minor injury around his jaw. He was thrown into the air and landed on his back.

Everyone was in total confusion as the surroundings were covered with dust. At first we thought the Hino was taken by landmines. Those who were trapped inside including me were removed.

I only had a mild concussion but no injury. I was later removed from the Hino. There was an old man with a bleeding nose. We found he had an internal injury which was caused by the barrel of an AK47. He was hooked by the barrel from the mouth before the pressure of the overloaded vehicle pushed it harder to tear the skin.

We slept in the bush before the UN truck coming from Chukudum pulled over near our vehicle. We put everything back in and by one pm, we drove to New Cush. We arrived at New Cush and spent three days and made it to Natinga next day.

We had to carry along the tobacco. There we were briefed by SPLA on the security between Natinga, New Cush, Newsite and Lokodak, because three days before we arrived there had been an attack on the United Nations convoy by Toposa herdsmen. Sometimes they do it to steal supplies.

We were stopped at the Natinga roadblock by the soldiers manning it. I talked to them before I slithered to new site and got another truck to Ladapal. Then the SPLA started forced recruitment and some of my friends were conscripted. I sold the tobacco and one gallon of oil given to me by SPLA officer Malou Ajak. I jumped onto a John Thor truck full of commercial cattle being transported to Nairobi for sale.

Before we left I saw Atem Deng being escorted by a solider. He told me that he had been arrested because he had a fight with someone on top of an Hino and they nearly fell off and so they were being disciplined. I asked the officer to allow me to take him for lunch. He agreed and came along with us. We ate together. I was then called by the lorry driver and I left.

By four pm, we had already made it to the Link compound in Kenya. Unlike before, I had a rooster and some money to sustain me in Kenya. Sitting near me at Link was a car mechanic. To those who knew him, he was a chicken thief. As I crossed the road to look for food, he had the opportunity to push the rooster into his tiny dusty room. He tied it in buckskin destined for export to international markets.

The car to Kakuma was waiting for me and I was worried that I had to let them go for I was still searching for the rooster. Lucky enough, a good Samaritan who was in the compound whispered to me that the rooster was with him. Immediately, I confronted him. He was boiling water to skin it. I removed it quickly and we took off. The following day, I took a minibus to Kakuma. As usual, I had to give the police what they illegally collected on the streets.

When I got back to group 50, it was already January 1998. With Imatong Primary school fitting in with my other commitments, I enrolled in class six. In my class were notable committed students such as Kuek Aleu Garang and Denis Dhieu Richard Achuoth. In the evening after group prayers at seven thirty pm, I would join, Achol Goch, to share a tiny tin full of paraffin as our lamp and together we would continue with our reading until ten pm against the backdrop of the year around Kakuma heat.

Achol was in form one. She was later married off. She was a committed young lady. She overcame all the odds of being a female student and had to double her efforts to complete her primary school certificate. She was also a devoted Jolwolic member of the Episcopal church of zones three and four.

In our school, we had Kuek, Bul Kunjok and Denis. These students were bookish. In our group, we had Chol Gai, Bul Yak Dau, and Deng Kuir Wach. Deng was my best friend and we could study together but he had the strength of being consistent. People like Bul Kunjok and Kuek

Aleu to mention a few in our class had the trick of non-stop reading. That was not me. When the result came, I would find myself pushed away from the possibility of getting closer to number four.

They had a vision that for them to excel, they had to get a scholarship. I was for any grade that would take me back to Sudan so that if I found a job with one of the United Nations Agencies, I could apply for job in the logistics section as one of the food distribution agents. These people were not badly off because they had to be provided with a few remaining food items as compensation for having dirtied their clothes with oil and flour all day. I came to learn that these students had tricks of studying in groups and at night. As usual, I did only two terms and then left for Sudan-Narus.

I joined the likes of Adoor Deng Ajang and Deng Kuot (Majith) in the goat selling business. We would move both cattle and goats between Narus and Kenya. In the evening two to three of us living near Bol Deng (Panaan)'s house would attend to listen to his music when he played his spring wooden guitar. He was a Red Army (Jesh Amher) singer and song composer. His father was nicknamed Der-angem for short sorghum which produces within short periods therefore assisting families to recover from starvation and stimulates the growth of buttocks.

School was not my top priority, though I had all the support from my uncle Abuoi who was like a father to me. In Narus, he strongly ordered that I should go back and concentrate on my studies for some time. He said education was the master key and that I should come back to SPLA later if I wanted.

On 21, October 1997 in Narus, I was sleeping. I woke up at four am and exited the hut to urinate. I could see around me flashes emanating from big torches. I thought someone was relaxing their nerves in the bush. It was not the case; these were SPLA soldiers having besieged their own town. They were looking for recruits. Then they knocked on the door. They wanted to force themselves in at five thirty am. I was called out from the hut. Then my back was hit with sticks by five soldiers.

I was ordered to remove my cloth. They collected enough recruits on the way before we were taken to the police prison across the road west of Narus. We numbered about seven hundred conscript plus three hundred volunteers who came from Kakuma. There was not enough space and thorns were used to surround and corralthe area. There were school buildings north of the Narus butchery and next to a deep crate created by bombs which later mixed cow meat with human flesh in 1996. Some of us were taken there.

I was with Adoor Deng, who trickly managed to sneak out between the frames of a small metallic window. His head was nearly stuck before we pushed it harder. He was limp, and for him to aide his escape, he started chasing children who were playing near the school. The soldiers who manned the gate heard the commotion. They came to us and nothing had happened was our reply.

The thorns were brought in by soldiers and an area of fifty by one hundred metres was fenced off. We slept there and the next day, we were driven to Kalacha near Kapoeta. Kapoeta is a strategic town which claimed the lives of hundreds of the Red Army and their commanding officers.

Mayom Atem Apei who was there on the fateful day when commander Majok was killed said that they infiltrated the town before they were pushed back by enemy fire. They lost the Commander but managed to repatriate his body. He was buried around the outskirts of Kapoeta and he said South Sudanese should name anything in his name.

We were to get basic training after which the high command would decide which towns we would attack. Inside the fence, we sang war songs common for the recruits of revolution. We did everything to ready our minds for any and all outcomes of the training. The SPLA commanders in charge of Chol Biar Ngang, Biar-Jaujau and Chol Alaak Ajak saw us off as we jumped into over six Hinos.

At Kalacha, we were fed on millet grains once a day. Our plates were hallowed wood used by Toposa cattle keepers to carry underground water during the summer to give to their cattle.

Some of us would feed on wild weeds. In the morning, we would wake up very early for a run. Our beddings were plaited bundles of grass put together to make something which looked like a mat. Twice a week we were able to shower. Any moves we made had to be properly supervised with more than five soldiers.

The way the soldiers were guiding us meant that they might have had information about frequent escapes of recruits from the training centres from their commanders. Commander Kuol Manyang Juuk was constantly heard complaining of recruits escaping to Kenya or internal displaced camps in neglect of their national duty.

Kuol's briefing was based on the consequences of being coward and betrayal of the people's cause. He hates acts of cowardice. He could easy slap the faint of hearts while standing in parade. My friend Atem who had been in Kakuma together with me in group 50 had big duck like feet. He was reported to have asked for permission to explain his flight to

Commander Kuol. While there he lied to him that he had swollen feet caused by a rare disease. He was not lucky. He was immediately detained and interrogated. It was found that he had been born like that.

Another guy called Andrew was allowed to go home. Andrew had a combination of sexual transmitted diseases (STDs) and his back was naturally deformed. He used to sleep near me about three metres from my designated sleeping floor. His back was not straight and was completely not fit for service. Atem with his big feet and solid body later became an artilleryman.

By eight am while having our parade, two strangers were brought by the SPLA covered with blood from the chest to the feet. They were paraded before us and we were warned that anyone acting against the requirements of recruitment would receive severe punishment like this duo.

Everyone got scared. However, I was not convinced for I knew about SPLA's war propaganda machinery. I wondered how someone with blood all over their body be running so fast with their own feet without collapsing. Deceptions were so commonly calculated in such way to achieve intended objectives. Deception worked so that half of the recruits believed that they were being taken to Addis Abbas for specialised training.

We were called and cautioned that anyone escaping from the training would be showered with bullets like them. We waited for the commanders in charge of training to give us orders for our next move. I left the centre with counterfeited permission. While walking in freedom from confinement and security escorts, I was indebted to people like late Gai Chol Awer, Bul Manyang Duot, Athou Bul (Gutakeu) among others. There was a possibility that they might have persuaded the commanders in charge to allow me to go back to school. These people were veterans of Koryom and their involvements might have helped in securing the tricky permission. Years later, I received devastating news of Gai and Bul. They were murdered by the Murle Militia who harassed Boma with the intention to dislodge the SPLA from the area. Officers like late Brig Gen. Chol Alaak Ajak of Shield One narrowly escaped.

During the nights that I would wonder if the Sudan liberation was our family enterprise or not. In our family for instance, it started with my brothers and logistical support, which was sufficient for the movement to reach its fifteen years by 1998. It was about five pm that I left the training centre.

I immediately without any form of hesitation crossed through the bush jumping here and there and finally got to the thick forest. On the path to Narus was Deng Garang from Palek. He was confused in the bush. He told me later that he was sent to collect firewood by his squad boss and thereafter he had an opportunity to take French leave (filer à l'anglaise).

I found him hiding between two trees looking like woodpecker. I spotted him, but I had to monitor his move to ascertain whether he was a mole following me.

I approached him and asked to join him. It was two brothers on the journey.

After two hours it was already getting dark. Deng was wearing a pair of shoes and was pounding the ground like an elephant. I told him that the sound attracted night orgies from a distance and that he needed to be mindful of his footprints as well.

Then we reached where Dhieu Yaak-matok was shot by the Toposa who were supported by Sudan Army to kill in return for guns. Two months earlier, I had been told that those caught by Toposa militia were sorted out by tribes while in Kapoeta. They would be interrogated to collect any crucial information on SPLA activities and then sent to the Juba military's deadliest white house confinements. Dhieu was ambushed by the militia after he left the training centre in the middle of the night. He was shot and he collapsed. They rushed to finish him. He gained consciousness and wrestled with one gunman. He managed to remove a magazine. He became unconscious thereafter because the blood was running over him. Then an argument ensued between his attackers.

When I met Dhieu a year later, he said one of the attackers wanted him shot in the head while the other ones considered such an action as a waste of a bullet on someone who was almost dead. They left him before the news reached SPLA at the nearby outpost. Ten soldiers were sent and he was taken to Narus. On that fateful day, Dhieu survived but one of the recruits went missing. He was reported to have been taken to Juba. We don't know what happened to him. I took five minutes to orient Deng on the tactics of walking in enemy prone jungle after we reached the hotspot. I had my own issue which was that I was wearing white trousers.

It was not changed for weeks because I was not given any chance to take my belongings on the night I was taken by soldiers. I had already folded it half way. My upper chest was naked to confuse the night raiders from identifying me.

At eleven pm, we heard the voice of Toposa armed men. There were six people properly armed. Immediately, I instructed Deng to make a five metre quick run into the bush and we acted quickly leaving the muddy road and quickly fell on the ground. I watched them and saw that they had one woman. I called Deng and proceeded. By two am, we had arrived at the school compound. It was the compound where the recruits were first detained before the next directions were given.

Then I proceeded to my uncle's compound. After one week, I proceeded to Natinga and New Cush. I spent the whole of May between the two outposts. Then the news of Commander Deng Aguang Atem Banylok's death had reached us in Natinga. It was devastating news having met him sometime back in Nauru. His death was at the Chukudum ranges at the hands of local paramilitary of the Sudan Army Forces. My uncle introduced me to him in Narus at the Sudan Medical Care compound.

He was renowned for subduing Sudan Army pro militants around Kapoeta.

I stayed in New Cush where I made a profound decision in my life. I left for and arrived in Kakuma and after a one week, I was approached by my aunty Nyandeng Kuir on the instructions of my maternal uncle. I was given some money to buy school stationery. With Akuol Kuir, we left in the morning to seek admission at the government owned Kakuma Primary School located in town. I took with me my class five certificate. It had my promotion to class six on it. Upon inspection by the head teacher, he ordered that I repeat class five. I refused and then trekked to Lopur at a distance of fifteen kilometres from the town.

I was admitted at Lopur together with Matiop Thuch Deng. I met the likes of Reng Gieu Reng (Rengo) and Diing Akoi Nyuon among other committed students.

We paid 500 Kenya shillings each for the whole year. The food at school was dried maize. The school regulation was that no food would be served to anyone who failed to provide firewood.

I shared a class with two smart boys named Ajok and Deng. These boys swapped the first position every time. Struggling with Kiswahili, I got pushed twice to position six. Then I was called by a tall one-eyed kind teacher who understood my challenges. Luckily, I rarely got flogged for indiscipline until I left Lopur.

I told him that Kiswahili was alien to me and that there was nothing I could do about it. I completed class six and I was promoted to class seven. There was insecurity in Kakuma as the Turkana tribesmen who lived around the region had complained about the UN and the refugees. They alleged that refugees were being fed at their expense. They had the assumption that resources had been given to the refugees and that as host, they should also get their token.

They began to come to the school at night and fire randomly. Our studies were interrupted. I was left with no option but to go back to the original school. It was already 1998. I completed two terms and then moved back to Sudan. I spent seven months there until January 1999. On 21 January 1999, I left Natinga with a truck but without a departure order. Then I reached the roadblock where I was removed from the truck by a soldier named Kudior manning the roadblock. He is related to my mom but was doing the SPLA restricted work. After an intervention from his senior, he allowed me to run after the truck.

The truck was going slowly because of the poor nature of the road. There were so many foxholes and spots. Then it passed under a big tree and I was thrown to the ground.

I fell which seriously resulted into minor injuries in the middle of my palms. I landed on my back and was just barely missed by the truck which was overloaded and pulling another big truck. It was a miracle.

The driver pretended to have not seen me because he did not want to take the blame. Also, he knew I had jumped on his truck without permission. I woke up from the ground and ran after him. He said nothing and by six pm we had reached Lokichiogo. I spent a night there and the next day I was in Kakuma.

The school was about to open. Those who were in class seven had got promoted to class eight. I was not; however, Kuer Dau Ngwei was still there.

Immediately, I went to him and spoke to him so he allowed me to move to class eight with conditions.

We had great teachers such as Malaak Ayuen. Then came the first semester, I took the first position followed by Denis. I scored 100% once in mathematics having learnt the mathematical techniques from my brother Mayen who was good at algebra.

We were about 112 candidates and the class was overcrowded. Come second term, I was pushed from number one to number five. It was because I stopped reading and that was a terrible mistake I made. As Kenya Certificate of Primary School candidates, we were encouraged to revise using previous national examination papers. Then Denis, Bul Kunjok, and I decided to sit for the mock exams at Sobat Primary school.

Only a few of us passed the exams.

Bul missed the first position in class because he was denied that chance by his cousin Dut Mayom who was the school flag bearer. On October 28, 1999, we started at Sobat centre. Around the examinations centre were police men moving inside three classrooms from one bench to the other to identify suspected cases of cheating. I completed examinations.

Those who got stung by scorpions on the scheduled day of examinations had to do their examinations while on sick hospital beds at the Kakuma Hospital. Those who were seriously sick with different type of diseases had to miss examinations for good or repeat classes. Sitting the examinations was must without taking into account any form of special considerations. Then I sat for the Kenya Certificate of Primary Education (K.C.P.E) in 1999.

By December, the results were released. Out of 111 because one student had passed on before the KCPE, only thirteen succeeded. Among us three had higher scores, but all of us missed out on a scholarship. In group 50A/B, there were only four of us namely Yaak Barach-magar, Deng Ajok Deng, Pajieth and I. The news of the increased number of successful students was celebrated as a group achievement.

The Group leader Kuol Ayuel Kotnyin was so proud that his advice to students was helping them make progress and that our promotion to secondary school would discourage idleness from those who refused to go to school. The group had young and old students and every day, there were many cultural activities from chasing girls to marriage cultural dances.

Once we were seen as achievers, beautiful girls who used to ignore us could give us powerful smiles. With three brothers ahead of me, distorting the line or corrupting the natural order of birth by getting married before them, or against right of turns could come with serious isolation or social control of being treated as an outcast from the entire family.

Deng and I would laugh with beautiful girls but with caution. At night, I would hear something like a lullaby. Unsurprisingly, those were sounds of poverty echoing from inside underneath my mat. There was no bed and mattress but a floorcovering. One could hear the collaboration between the hard surfaces with exposed ribs when someone is straightening oneself on those floorcoverings.

Then any planned further interactions with many beautiful girls would disappear. In the village, I had left a decade ago there was only uninhabited cow byre and abandoned farmland. I had nothing but an

United Nations High Commission for Refugees card.

Those were not assets to convince any girl to be my wife. There was nothing to rely on to persuade any pretty beautiful woman of my age. Poverty in the camp and even in the rebel areas made annoying recurring sad sounds in my mind like crying babies.

I also knew that I should not ruin my education with short term human desires and happiness even if many desperate girls in the camps had no issue with poverty but love. Love does not bring the food on the table were the words of by community elders, later echoed when I was in Form One by Yaak Deng Akoy. Yaak said that people do not graduate from it. It has no university of its own though one can study it as a course.

We celebrated the end of 1999 and received 2000 with excitement. However, there was an assumption that it was going to be the end of the world because some people had lied about it. In the wake of 2000, we were accepted to the only available secondary schools of Kakuma and Napata.

I was admitted in Form One in 2000 at Kakuma where I studied until the third semester of 2001. But before the end of year examinations, I went to Eldoret for the first time. This was because my brother Mayen had resettled in the United States as of 21 June 2001. To leave the camp, one must have travel documents. So I took an UNCHR travel document and the reason was that my brother who was one of the lost boys and settled in the USA under humanitarian visa had requested that I visit him.

As usual, I took the minibus commonly known as matatu. The motorists did not leave in the day until seven pm. This enabled the police to extort money.

For instance, the first police roadblock removed all of us including the sick and vision impaired person (VIP) from the bus. I was in charge of three women whose children who had settled in the USA and had begun to send her money. They were heading to Nairobi but they had to rely on me to break the language barrier.

The policemen were looking for minimum of 250 Kenya shillings or higher from each person. Sometime the driver would collect them and hand them over to the police or have everyone talk to the police directly. The search was purely directed against refugees but not Kenya citizens. I collected the money for the three women including mine. It totalled KSH 1000 collected in less than ten minutes. This was huge amount for those whose salaries were below KSH 5,000. The police accused me of not having done a proper job and he punched me putting pressure on my eardrum.

By the time I arrived in Eldoret on 15 August 2001, there was already an internal injury on my left ear. The policeman was using his right hand to abuse the right to protect the vulnerable groups giving to him by the state.

Dinka refugees called the police 'bolachol'. These are two words combined. For instance, Bol in Dinka is boy following the twins. Achol is a female name either taken from grandma or derived from the compensation for the loss of a beloved one in the family. Whoever coined the word had their reasons. One of the reasons might have been because most of police officers were eating from many sources from their statutory mandated salary to extortions of money through threats. So the first and second rounds were all going into their pockets.

The use of one's position for gain is corruption. Corruption paints the inside life of its beneficiaries very dark. Shining out is material wealth from the general view of the public, but on the inside the corrupted are totally covered in murkiness.

The police were doing an amazing job too such as protecting travellers from night robbers. In the camps, the police maintained law and order since cases of communal fighting were prevalent. Some cases such as impregnated girls, adultery, fighting over beautiful girls could rock the camp any time and therefore police presence was crucial.

Bribery is like floating vegetation which cannot properly support desperate drowning men. It is like the grass in the rainy season before the Sudan African summer heat erases it completely. On the way from Lodwar to Kanok, we went through a lot of police checkpoints, and interaction. They would ask we support our travel documents with money.

By eight am, I arrived at Eldoret Town. Above me were tall buildings. I was looking up like a chicken ogling flying predator eagles. Then I took another bus to Kapsoya where I was taken to my maternal Uncle Abuoi's house. He was renting a small house in Eldoret having been sent there on the permission of Dr. John Garang de Mabior to study a Bachelor of Medical Sciences/Medicines at Moi University.

The next day, I woke up very early. Then I proceeded to the house of Most Right Rev. Bishop Nathaniel Garang Anyieth Jangdit.

His Lordship has been our bishop during the tough days in the rebel held territories. He was a bishop of liberation and sometimes talks of his time in Anyanya One. He is righteous man who speaks wisdom and blessings.

Mobile phones were not commonly available those days. They were limited and also not many people were able to afford them. In Kakuma, I used to communicate with my brother Mayen through letters brought to the community center located near the zone 3 church by Kenya Red Cross International.

Inside Bishop House were mobile and office phones. By the time I arrived near his compound, there were already over fifty people at the gate waiting to be called. The Bishop family had generously made their house like a feeding centre. We stood outside on beautiful evergreen grass. There was already peace in the evergreen and beautiful country Kenya. We continued to wait as the caller had to first identify who wants to talk to whom.

There were thousands of lost boys/girls scattered all throughout the United States. All of them wanted to inform their relatives that they had made it safely and were willing to send bundles of dollars. The phones at the bishop's house had only two numbers distributed to many lost boys or girls. Whenever a call was made, there would be no privacy. For instance, every form of one to one telephone conversations from internal family affairs, marriage, all sorts of ailments, abortions and money theft would be for everyone's consumption. Those who had cattle in the villages would stretch their arms higher in imitating the curves of their favourite bulls with the assumption that the beloved ones in America would know the progress of growing bulls.

At seven thirty, the phone rang. The caller was Mayen. Then it was my turn. Since it was the first time I was using the phone, I started with hello, hello, hello as the volume of his voice became even louder. Then his voice became clearer and we spoke about school, and how to support family members in the refugee camps.

If at all I had disclosed anything confidential during our conversation, then I left it to the discretion of those who would be sipping on bottles of beer in Eldoret Town scorning how people are not used to having telephone conversations.

Mayen was based in Tucson, Arizona. He sent me $400 US dollars which was equivalent to over 30,000 Kenya shillings. In the morning, I rushed to Western Union with all the details.

It was a lot of money and I moved like a rich man in town. Now, I had to celebrate. First, I took my friends to a restaurant where we ordered Nyama Choma (roasted goat meat).

In the refugee camps, our yearly diet was lentils and maize. Refugees rarely got beans at Ahoyou and Nubian small hotels.

With Nyama Choma, a proper life was about to show its signs. It had begun right there at the bank gate. I bought clothes and I thought of a bright future. Then I spent two months and on the 10 September 2001, I left Eldoret for Kakuma. Some of my friends such as Dhieu Ahou Dual and relative Chol Garang Juach had already sent me their financial support. There were also instructions from Mayen for me to find a good school in around Eldoret or Nakuru in accordance with my uncle's early advice to him.

Before I arrived in Lodwar, there was news of the September 11 terrorist attack.

It was shocking and devastating for we had relatives who had just arrived in the USA leaving behind the horrors of war in Sudan only to find ourselves watching other evils. I read it in the newspapers that the terrorist had killed thousands of people and destroyed properties and that this had triggered the USA's long time hidden intention to invade oil rich Iraq on 19 March 2003.

By nine am, I arrived in Kakuma town with a big bag. I hired the only taxi in town, the bicycle to group 50B. Then everyone was coming to me inquiring about the conditions of those in the USA.

Inside the middle of compound, was a dinner bench made of wood. It was where we ate together and sometimes got advice from people like Mawut Yak Duot and Thiong Dabek (Yar). It was our dinner sitting place where Mawut would pay a short visit and encourage us to study harder knowing that education leads to meaningful living. I sat the secondary school third term examinations and then I was promoted to Form Three. My performance was dropping and teacher Machar Buol was not happy. Having been a long-time friend of my brother Mayen and having had constant competition in class with Mayen, he was a recent A- student. He was providing extra assistance to Kakuma refugee secondary students.

I had done my first term in Form Three before Mayen sent money so I went back to Eldoret during the holiday. It was my second time in the Rift Valley and this time I used the bus compared to the previous time I had travelled along it with the lorry carrying goods. I arrived the next day and proceeded to Kapsoya. It was already January and schools were opening.

I started looking for a good school. I went to Arnesen High at Burnt Forest on the road to Nakura. I was surrounded by native students who were not familiar with smooth dark skinned Sudanese people. I could hear some susurrations with respect to my height and size.

After fifteen minutes, I was asked by head teacher Kimathi Rurii to presence my certificates. He perused them only to report to me that there was no vacancy in Form Three. Unknown to me was that Form Two had Thon Deng Dhel. He was an intelligent gentleman. I left the school and proceeded to Nakura. I slept at Abul Deng Reng's house before I went to St. Columbus High School. It was a mixed school of handsome boys and beautiful girls.

I did the first term and got the second position in class. Ali, an Islamic religion oriented student was number one. He could rarely greet women although I challenged him as to why he greeted his mother. His view on women was terribly negative.

In class one where Bol Garang Bol and Adoor Wal Biar were was a Kenyan-Somali born student named Hassan. He used to come to me with the Quran in the evening before we broke for dinner. He would discuss verses about God and other prophets that were in the Old Testament. He would say that if Jesus was the son of God, He should not have called his father to help him out when he was in trouble in the sea and on the crucifixion day. He would add if He comes back he would first start by blaming Christians and Jews for neglecting worshipping his father. He concluded that Jesus was not crucified, but Judas Iscariot was in his position because God had taken him before the mop came. But the funny part is that he had massive knowledge of the New Testament. When I decided to take the Quran from him, he refused. He said that I should not touch it, because I am faithless. I laughed at him. He was a good person who could provide me with his plate when lunch or dinner time arrived.

In the Sudan I had left, we had Islamic extremists who advanced the spread of Islam with impunity. Ali was a committed student not like moderate Ahmed who would get busy at break time kissing one of his girlfriends on the lawn yard behind Form Three.

The school was a day school where students come in the morning and go back home in the evening. I was living about twenty kilometres from the town with the beautiful children of my aunty Abul. It was in the dusty Langalanga suburb after the Deliverance Church to the North West. So I had to take the bus to town and then to school. I was concerned with my school performance during the national examinations and the fact that I took the second position when I was getting very low marks in Kiswahili was like sitting in my house of self-deception.

There were also sensual looking beautiful school girls nice from head to toe. Those girls were the light of Kenya's diverse multidimensional society. They could shut down one's neuron system for few seconds. Their passing nearly makes one crack a pen with teeth. So the school was not my choice.

Then we broke off for a holiday. And instead, I opted to try to get into Arnesen for a second time. Arnesen is one of the top secondary schools in Eldoret East Constituency which had produced the best Kenya leaders since its inception. This time I had to go straight to the head teacher Kimathi.

I explained my desire to study in his school. I showed him my class certificate. He accepted and I was provided with an admission form. However, I had another decision and that was to repeat Form Two to get a good grade in the Kenya Certificate of Secondary Education (K.C.S.E).

Unlike in Narus where I was running temporary businesses like buying and selling goats, I made a choice to put my life in the correct perspective. I decided to aim to be a lawyer. I left the issue of AK47s behind because I watched Kenya television and was inspired by Kenya's Attorney General Amos Wako and Yash Pal Ghai.

It was also not Kakuma where hundreds of committed students were looking for merit based scholarships were relying on a dim light stemming from small tins filled with paraffin before they got scholarship to various schools in Kenya's major cities. These tins were later replaced with special lamps which were not affordable to many students in Kakuma.

To overcome the urge of students engaging in sexual activities, there was unproved speculation that some schools were adulterating the food with a small quantity of paraffin to suppress any form of sex drive that the students might have had. In the different and cold environment of Eldoret, there was electricity throughout. It was not like the days where I used to rely on paraffin to prepare my homework and revise for my examinations.

I met trained teachers and the teaching was excellent. The school was a boarding school for only boys. I met Thon Deng Dhel and at our organised dinner table, we had a Nandi student named Brian who became our comrade. We could share everything because the ideal of socialism had impacted in our life in the bush, rebel areas, and internal displaced and refugee camps. The union of sharing scarce resources made us look healthy and that caused jealousy among the native students. Some thought we were fed very well by the school. Others were also not happy with our performance as people like Agou Anyieth Kur and Bul Manyok and Deng Garang were taking the top positions.

The school was very proud of the committed spirit of people as some of them were former soldiers. For instance, Deng was trained soldier who fought before he left Anyidi in 1999 under a kind hearted Israeli lady called Doreen. First she proposed that Deng should go to the Maridi training centre to study polytechnic courses but it was agreed that he should move to Eldoret where he would stay at his uncle's house during the holiday. There was no doubt some of us were also dealing with sciences and mathematics to the best of our ability.

Brian was a very good and God fearing person. He was so kind that sometimes he would take us to his village to spend few days. We would eat nice Nandi food and learn about Nandi in his native humble grass thatched houses. The development of African societies were visible only on the roadside and major cities while the life at peripheries was so shocking. For instance, there were people still taking water from the boreholes in Kenya after five decades of its independence.

In term three, the school had a squad: Deng Garang Bul, Dhuol Mach Dhuol, Bul Manyok Duot, Thuch Garang, Agou Anyieth, Paul Mac Awer, Kut Alier Apollo, Thon Deng, Juuk Ayor and me.

We were ten in number. John had one of his fingers broken after they assaulted the Sudan Army before he left for his study. He was decent person who was committed to the freedom of himself and his country in the future despite the terror of war. He would tell us that he had done his part in the liberation of Sudan and later South Sudan independence.

Students like Bul and Thuch were Form One students at the time while the late Kut was still finishing Primary Eight at a prestigious school. Like the rest of Form One's new students, they were referred to as Mono.

Figure 10 from right Juuk Ayor, Bul Manyok, Dhuol Mach, Thuch Garang and I and Unamed Bul Kenyan friend at Nakura falls near Nakuru National Game Park 2003

Figure 11 waving fifth South Sudan Independence flag on Dr John Garang Mausoleum as we received his H.E Gen Salva Kiir Mayardit, South Sudan's President, on 9 July 2015 from VIP platform

Anyone in Form One was regarded as molecular (monolayer). It was mono responsibility to clean the dormitories and school compounds and mess. People like Thuch who dubbed his name as bull of all bulls (*Moremamor*) would protest for he believed that parents paid school fees for their studies and that the school administration had no right to make them workers. However, it was not the Arnesen that was the only school doing it. The majority of schools were practising it either illegally or approved. Those who refused to work received varying degrees of punishment from the head-boys.

South of our dormitory was the school farm and dairy cattle. The farm was used to cultivate maize which was later used to feed the school. The dairy cattle were providing milk to use as our watery milk tea. School

Figure 12 Late Kut with an unidentified white man and Kenya police officers at the background before he was innocently murdered in Malakal 2017, South Sudan

discipline was stimulating students' competition. Students such as Peter Kariki from a Kikuyu was our class top achiever. His eyes were always glued to the books. He rarely spared any available time.

As a provincial school, we had to take part in the sciences and mathematics congress with Moi Girls High School (Eldoret).

Whenever, we met the Moi girls, we found them shining from head to toe. When we got back to school, our mathematics teacher would challenge us to be equal to those girls which always toppd Eldoret East Constituency. Motivated by desire to succeed academically, Arnesen began to receive a number of Form Three and Four repeaters who badly wanted to get government grants and scholarships abroad.

There was competition at Arnesen. As a result, I was pushed away from the top ten and even further to the twentieth. I was doing poorly in Kiswahili despite the attempts by the school administration to have the teacher concerned allocate time for my extra tutorial assistance.

Instead, I put more effort into all the other subjects. I became creative in mathematics and Bul and I took Bul-Jok mother of Construction Theory to the national mathematics competition. Our project was how to find any degree with a ruler, pencil and protractor. We could easily bisect and find all forms of degrees based on your choice. The science teacher was impressed and the project was approved to go through to the district competition.

One bright sunny day on 10 July 2003 at the Uasin Gishu Secondary school complex, we were knocked out by the Chesogor Secondary school. We took our school bus back to Burnt forest.

We found our usual feeding of Sukuma wiki. Sometime we could eat meat three times a week. Sukuma wiki was eaten together with Ugali, which is a type of solid maize flour porridge.

I would hate boys' school on the day of meat. Giant boys nicknamed Somali and Kataamism would nearly crash Form One students. The beef was not enough for over 750 students. The Form One students were aged fourteen to sixteen. One day, Kataamism dipped himself inside the porridge in the process of fetching two cups more than any student. He was ashamed. And whenever students saw him passing, they would joke behind his back that he was practising the agricultural session of plunge dipping of livestock.

The school had a rabbi. We also had a television embedded inside a frame in the students' mess. Church prayers were compulsory for everyone either early in the morning or evening.

While at Arnesen, I completed Form Two and got promoted to Form Three. The four form students of 2003 were not committed save for a few.

The school administration was expecting more commitment from them but some were practising the arts of marijuana (bangi) smoking. Some students also had issues of dishonesty and lack of focus.

Then one Sunday, a Form Four boy concealed himself under the bed inside the dormitory named Mountain Kilimanjaro. There were four dormitories with each named after prominent Africa mountains to aid the geography students.

The student was out of sight of the teacher on duty to attend to the school service. The teacher name which is a variation of Malachai in the bible and had angelic meaning was in charge of supervision to see who was violating the church service compulsory attendance regulation. He was assigned to ensure students attended service every Sunday. That day we had service by pastor from the Deliverance Church. He came from Eldoret and his service was to end at eleven am. Then everyone went to wash their clothes. It was already evening and everyone attended the evening reading.

Unknown to the teachers, was that a certain student had decided to steal the school's only TV. He immediately sneaked into the students' mess at night and removed it with ease. In the morning, as we ran for our morning tea served in the same hall, there was no TV. The school had guards by the gate but fence hangs as a torn skeleton of materials.

It was the only television we used to watch the world news from the American global war on terror, or the world cup to other extracurricular activities from around Africa and beyond.

His plan succeeded. He removed the TV unseen. By the morning, news had spread quickly. As a result, the school administration organised serious prayers and intervention powers of natural beings were invoked. The pastor prayed that the powers of the Heavenly Father would reveal the thief. Such an invocation of all the powers in heaven made him come forward. He confessed and cried in front of everyone.

Kimathi was a head teacher of high calibre. He asked that the student be forgiven by his colleagues and teachers. Then he was forgiven, but his punishment for engaging in illegal activities of no meaningful contribution to his future was an expulsion. I imagined how his poor parents would feel when their son finds them sweating tiling a land to plant maize, cabbage and sukuma wiki.

Every Saturday, the school invited inspirational leaders from across many disciplines in Kenya namely, the Church, universities, and the medical and law professions. The former lawbreakers and HIV (AIDs) victims would narrate to us their experiences and touching life experiences.

One of the notable speakers who held a Masters in Psychology would make us laugh our lungs out with hilarious stories such as an injudicious student illegally sucking the school's only milking cow's teats at night. That his jaw was kicked away by the cow was attention-grabbing. Then he would conclude that the consequences were serious and disgraceful to him and his parents. He could not explain how he hurt his jaw during the examinations time.

All the hustlers of the modern Kenya who made to the top leadership in the country would come to school to give free lectures.

Then the talk would end with us being rewarded by school administration with three pieces of bread each. The day would be concluded with inspiring stories. On one occasion, I was awarded the certificate of Merit and Excellence Efforts for one of the poems I recited. It was an imagination of school bus waking up in the morning to drink the student's porridge before the policeman at the gate caught it red handed. The stories had to have both fictional and non-fictional elements.

The last notable speaker to come to our school was the Vice President of Kenya Moody Awori in 2003. I was about to recite a poem on the Sudanese civil war and its associated suffering but I was unprepared. We paraded in the school compound about a hundred yards from teachers' office. His Excellency was welcomed with gospel songs and other activities were run before he finally addressed the school.

He left thereafter and I suspected he might have been invited by school as part of its fundraising to buy a school bus which students later called Malaya (or prostitute). To raise some money, the school administration used to hire out the bus to people who had a function. Some students with nothing to do other than taking drugs did not like it. They even organised a students' strike complaining about us, their pastor and catering team and the bus. They argued that the bus was not prostitute and that it should stop. Notable politician Tinderet MP Henry was there busy interacting with students.

Kenyan politics was so interesting and when elections were about to approach you find politicians intermingling with ordinary farmers at Kapsoya's small and dusty shopping centre. As a result, I developed adoration for Arnesen's surrounding environs.

We also learnt about Nigeria's societal issues in those days of enormous military coups from Chinua Achebe's short story of a corrupt minister of culture, in A Man of the People. The school did everything to see us excel.

We were provided with all the necessary resources and with an assurance that teachers would complete all the subjects before September. By the end of October, 2004, we started our Kenya Certificate of School Education (K. C.S.E). These are the standard national examinations in Kenya sat at the same time all over the country.

On 11 November, we had finished the examinations marking the end of twelve years of academics. Previously we were given opportunities to tour the Moi and Egerton Universities including the precious stone decorated home of Lord Egerton. We were once taken to Kerich tea plantation. The Nandi hills were occupied white settlers' lands before Kenya's independence.

In History, we were told of Africa's resistance to Whiteman rule on the continent. Then we read about Nandi resistance from 1890 to 1906. Their leaders were Koitalele Arap Samoei and Orkoiyot. Informed in advance by their prophets of the coming of white human beings on their lands and th suffering associated with occupations, the Nandi hills became fighting battlefields until their leaders were murdered by British Army Officer. It was a cowardly act at a time when peace was accorded between the Nandi and the white settlers.

We were allowed to pick the tea leaves. I took one and chewed it. Kenyan tea has a nice smell. Deep inside the planation, we found hundreds of tea workers busy packing the green tea. Then we headed to Moi University's Kapsabet Camp. There was one powerful pictorial sign at the entrance: "Academic Highway." I made a serious note of it before we proceeded to the Department of Engineering where we met Prof Ajuou Magot and Engineering student Deng Diar Manyok. We were shown how distilled water was processed.

This was meant to inspire us to look ahead to our future. Back in Arnesen after examinations, the reactions of some students were to burn some of the papers and exercise books of the subjects we loathed. Mine was Kiswahili. Others were going for subjects of their choice. Talents are God given and different and only a few people are jacks of all trades.

While I was doing this, Kiswahili, was slashing my marks and lowering my grade in the Nairobi marking centre. It was tit for tat once considered by Williams Shakespeare as capable of turning the whole world blind.

I packed all my belongings and headed to Nyeri District to the small town of Karatina. It was where my family had relocated to. We knew it was the end of life in secondary school and the world of Kusuma wiki and ugali. While at Karatina, I developed a habit of going to the Karatina county library to read newspapers on issues around the globe.

There were Sudanese Peace Talks in Naivasha, Nakuru, Kenya and Nairobi. The Sudan government of the Islamic National Front with leader Omar Hussien El Bashir delegation was negotiating with the Sudan Peoples' Liberation Movement/Army led by Dr John Garang De Mabior. Children were born into it and see either exposure to war or displacement. The newspaper reported a compromise on the right of self determination and popular consultation for the peoples in the Nubian Mountains and Blue Nile.

By February 9, 2005, the Comprehensive Peace Agreement was inked and a grand ceremony attended by hundreds of thousands and hundreds of dignitaries was organized in Nairobi, Kenya. The news reached us and I left with late Majok Atem from Karatina very early in the morning. The ceremony was attended by hundreds of thousands of people. There were enormous groups of traditional people mainly from all the tribes of Sudan.

Figure 13 Bol Mach Nyikuany (centre) and late Majok Atem Mabior June 2010, Juba, South Sudan

Aguil Chut Deng, who came all the way from Australia and a member of the liberation council organised a fantastic cultural Dinka group. I just arrived after two hours' drive and were among those who were loitering at Nyayo stadium before the Kenya police asked that we move out of the stadium and find somewhere to sit. The stadium was full to its capacity.

Then we were alertd to get ready to receive the dignitaries from all walks of life from the United Nations Secretary General Kofi Annan to Kenya's President Daniel Arop Moi.

His country was the host of the peace talks and was instrumental in ending the civil war through six protocols from a referendum for an Autonomous Southern Sudan, Abyei to popular consultations for Nuba Mountains and Blue Nile.

By nine am, the ground was cleared for the arrival of dignitaries from Kenya President Daniel Arop Moi, Omar Hassan El Bashir, Gen Sumeiyo, to Gen. Collin Polin the United States' special envoy to Sudan who helped the parties to make a breakthrough during the peace negotiation during the George Walker Bush Jr Administration.

The numbers of foreign dignitaries were so many that the majority of them could not get a chance to speak. In the unusual response by the mammoth crowd at the stadium receiving Dr. John Garang De Mabior, we almost shook Nairobi with our feet. We were shaking everything like an earthquake. The joy of happiness was deafening to the dignitaries. Dr. Garang was walking closer to Beshir who was known in marginalised Sudan by millions as a murderer.

In Kenya, he had few supporters in the stadium not like Dr. Garang. Garang as an Africanist and people's revolutionary leader was on the soil of his Africa. Garang whom the protocols made Sudan Vice President and President of the government in the south was cheered on like a messiah of poverty redemption by Kenyans and millions of people and television viewers inside and outside Kenya. He was a rare charismatic leader who had drawn millions of people to his idea of a new Sudan.

With the marginalised being the majority in the Sudan, he was the people's redeemer. The ceremony started with prayers by the Kenyan bishop and immediately the rest of procession began. Keynote speakers narrated the turbulence with Sudan since its independence and continuous bad governance. Then Dr. Garang was given a floor and we stood up straight and sang any available songs from church to the revolutionary (war) songs. He started his speech with dignitaries before touching our hearts that we had suffered in dignified silence for so long with our women who walked several miles in the day and evening to look for five litres of water. He continuously underscored of human suffering from wars to natural disasters. He was remorseful for the level of destruction and deaths the 2005 Tsunami had caused along the coastal regions. He spoke for nearly half an hour until he confessed that since he was a guerrilla leader he therefore had the right to speak for several hours. He prophesied that he has seen his people taking a thousand years to smash themselves into history.

In the midst of his speech, he confided that there would be no more bombs and that Sudan would see the happiness of children and women. I was very close to the very man I had missed crossing during the army struggle. He was in Maridi briefly when I had gone to school and Nairobi was great opportunity to note every word he spoke.

Courtesy of Sudan. Net- Figure 14 John Garang statements on 9 February 2005

Garang was a rare African leader. His humourous character used to motivate the guerrilla fighters who fought without salaries. He was the engine to propel the movement. He acknowledged the deliberate disappearance of the ancient kingdom of Merowe and among others sultanates and Nubian Coptic Christianity of 15th century. He prophesied that the National Congress Party of Omar Hassan El Beshir was too deformed to be reformed.

As he spoke, I stood about forty metres from where he sat. He sat two to three rows away from Omar. Then I directed my eyes to Beshir. He was not really in a good mood as his dark file was being exposed with hard facts to the whole world. Instead of looking straight to Garang, he was looking down and squirming. Sometimes, he pretended to be beating his bullet deformed leg with his stick.

He was outsmarted by Dr. Garang. As the people's leader, Garang ordered the release of prisoners of war. In our days in Natinga, we had hundreds of prisoners of war making a living in the community as beds and tea makers. Those were the people he was ordering be set free. Bashir who had spoken earlier had nothing to release. Our men like a friend of ours whom we lost on the way to Kalacha were murdered. There were no international humanitarian laws in Bashir's reign of terror. He was an unremorseful Islamist operating on the strict sharia code as interpreted by his godfather, Hassan El Turabi.

By four pm, I left with Ajak because we lived far away from Nairobi. The crowd was also dispersing. And the next phase of the implementation of the peace accord with a security agreement, was the deployment of forces along the lines of the South-North border as it stood on 1 January 1956 and included the swearing in of Dr. Garang.

While I concentrated on my resettlement in Australia, Sudan was already five months into permanent cessation of hostility. Arrangements were made and the briefing from Commander Chagai Atem at the Yaya Centre home of Dr. Dau was that John Garang would go to Khartoum. He said that they had security concerns because the SPLA unit stationed in Khartoum was not enough. However, he confided that Garang would go and be sworn in July 2005 as the Vice President of Sudan and President in the South.

He would remain commander in Chief (C-in-C) of the SPLA and other organised forces to be restructured from the mother SPLA in accordance with the Interim constitutions governing Sudan like two entities. He arrived in Khartoum after an advance team of people like Pagan Amum and his security team had earlier made it.

Something never seen in Khartoum happened that day. Millions of people dotted the square like ants to receive him. Those people might have included the parents of children the President sent to the battlefields in the South who never made it back and those who had suffered under Sharia law which denied every person any forms of freedom to the extent that people were resorting to wine stealing to keep themselves happy. He was sworn in and spent less than three weeks there.

During the armed struggle, Garang had friends who stood by him and those who later became his bitter enemy. President Museveni of Uganda, a man who had joined with Garang on common interest and mutual understanding helping each other during the terror of the Lord's Resistance Army was his great friend and ally.

He was presenting during the signing of the peace agreement and was referring bitterly to Beshir and Arabs' policies of trying to create Arab states in the African Continent. He was on that occasion repeating Garang's remarks on the problem of the Sudan to build a pure Arab state without Africans.

In unclear circumstances, Garang left for Uganda but he had to pass through his base at New Site. The late Atem Aguang Atem was close officer to John Garang. He told me how they were instructed by Dr John Garang to mobilise troops to move to Juba. He said they were instructed on the evening few days before the end of July 2005 to put together all the artilleries and heavy weapons. That was the last meeting he had with his Commander in Chief. He said sitting outside waiting to see John Garang was Clement Wani. He was in New Site on the order of the chairman. He had instructions to receive his brothers and that the war was over. He was part of those those who had countered the SPLA offensive in Juba 1992.

Having given instructions and orders to be executed by his senior officers, Garang was flown on 29 July 2005 to Uganda with a chartered plane from Kenya. He was in Uganda to meet his long-time friend Museveni. He had a meeting with the President of Uganda and spent a night there. The next day, he was given the Presidential helicopter M.I 175. However, he was flown in the evening. His helicopter was heard around the small hills of New Cush bypassing a mile away from SPLA barrack where were waiting for any orders from the leader on the implementation of peace agreement. Unknown to his soldiers the helicopter was the one which claimed the life of our beloved leader trying to make a turn. After a few minutes, there was explosive like sound according to Atem who was playing a game at the time and among the rescue team to recover the wreckage and bodies all the passengers including Sudan Vice President. There was a barrack east of New Cush, which I remembered vividly in 1998 having spent two months at Pastor Deng Lou home. I was in Karatina waiting for Kenya's news.

Then came the breaking news by Kenya Televsion Networks (KTN) of the disappearance of SPLM/A leader Dr. John Garang. The reporter began narrating how the helicopter disappeared and the possibilities of other theories on what could be the cause.

This report was that Garang's phone had been ringing but no one picked it up. The news was the saddest. And I left the dinner table instantaneously. My food made of ugali and cabbage was left cold. I rushed to my room. Tears were running down my cheeks. Then I switched on the radio searching for alternate theories to the disappearance since statistics on any plane which lost contact always ended in tragic irreparable loss.

I was waiting for the worse. Garang was not only son of Wangulei who had climbed a rare ladder since the era of Abel Alier and Joseph Lagu, but also the brains behind the success of the movement.

Our hopes either in the refugee, or internally displaced camps, and warzones were based on the man who had left a comfortable life in Iowa to shelve his Doctor of Philosophy Degree to liberate his enslaved people with AK47s. Garang had on many occasions had tried to create awareness for our people to know their rights. He was not happy with those with weak hearts.

For instance he had to challenge people in his speech in Rumbek after the fall of Yei to the SPLA in 1997 as being feeble in their hearts. He was like Piou Niop de Jesh Shaba. He had overcome the enormous challenges of managing a guerrilla movement driven only by motivation and resilience. In all the darkness and hours of great agony, people need leaders. He was one of those leaders.

I remembered in our Sunday Service in Natinga along the Sudan-Kenya borderline, one of our church members read to us about the plot against Jesus in Mathew Chapter 26. He was reading that when Jesus had finished saying all these things, he said to his disciples, as you know, the Passover is two days away-and the son of man will be handed over to be crucified. And while they were eating, Jesus took bread, and when he had given thanks, he broke it and gave it to his disciples saying, take and eat; this is my body; and did this for cup of wine and said drink from it, all of you. This is my blood of the covenant, which is poured out for many for the forgiveness of sins. I tell you, I will not drink from this fruit of the vine from now on until that day when I drink it new with you in my father kingdom. And that the spirit of his sleepy disciples is strong, but that the flesh is weak. Rev. Deng came to the same conclusion about the rebel movement.

Few years later when Dr. John Garang was in his ancestral home town of Bor in his tour of southern Sudan to disseminate Machakos' protocols of the comprehensive peace agreement, he had to narrate a proverb about a Dinka man who could not eat the meat of the bull he kills as Riang, to be called by the name (Agutmayom) of that bull. Riang is a ritual practice where an able man would spear a bull which is later finished off by the colleagues and eaten in his total absence.

The person is culturally prohibited from even using the plates used to eat the meat. It was taboo as anyone who eats from his bull would go completely insane. Garang was created by the creator with purpose like other leaders who had foreseen their death such as black American liberator Dr. Martin Luther King who saw his death during his 'I have a dream' speech of 28 August 1963.

In his days in the bush he would refer to the surrender of Arab fighters in hundreds during the Jungle Operations Storm (JOS) and made jokes out of it. He made a mocking statement in Rumbek that the Sudan People's Liberation Army had never seen a mass exodus surrender of Sudan Army Forces (SAF) recorded in 1997, because what they had got used to were the surrendering of some of the SPLM/A high command commanders.

Eight hours during the night passed like one hour. I was not sleeping. The next morning would present a clearer picture of what transpired on 30 July 2005. Then there was BBC news at around four am. It had the satellite relocation of the wreckage at around Kidepo ranges south west of New Cush.

The news was shocking and many people included John Deng Garang who received the news in Australia state of Queensland were admitted to hospital for mild nervous shock. He recovered only after a few days. He was a child soldier. Bul was among the men who left northern Kenya Kakuma in the midst 1993 to join as volunteer fighters.

He was trained at Lotuke and put under the command of the former SPLA Affairs Minister who died decades later in a plane crash, the late Gen. Dominic Dim Deng. Bul might have gone through the hardest struggle for freedom and saw a bleak future without Dr. Garang.

Some inpatients had to later pass on. The news was unbearable. The news of Garang's death reached Khartoum causing violence and riots never ever seen before. The international and local newspapers reported over eighty deaths, burnt cars and properties. Hundreds were also wounded. The mistrust between the successive governments in the Sudan and southern region is as old as the time of the slave trades which nearly depopulated African societies of their best brains and human capital.

Other available records of the assassinations of African leaders were part of grievances taught during the military training. War songs with mentions of past leaders murdered by Arabs fed our liberation.

The death of Dr. Garang was considered as an assassination. Dr. Garang had friends and enemies. For instance, in a small lane called Wimbi next to the 680 building at Kenyatta Avenue in Nairobi, sat an enormous number of jobless Southern Sudanese. Some of those would look on to those dressed in nice suits and black shiny shoes bypassers and started gossiping. In the midst of Wimbi near I.M Building a few busy individuals were circulating booklets in Nairobi. The contents of those booklets were anti Garang. Some were throwing words that said that the exodus era Israelites leader Messiah Moses did not make it to the promised Land. Those were compromised southern Sudanese for Nairobi had the Sudan Embassy which was doing everything from bribery to encourage subversiveness against the movement. A man called Ali Haj working with Sudanese government was known for bribing high ranking officers to support acts of sabotage against the southern movement.

However, Dr. John Garang De Mabior knew since day one that the journey he took was the perilous route. To liberate people whose blood had been the source of life for their oppressors was probably life or death mission for him. Either way, he was prepared for anything as exemplified in his speech at the fall of Torit in 1989. He began with saying that he hated oppression and was willing to shed his blood to liberate the poor masses.

Hundreds of miles from Nairobi were violent riots in Khartoum which claimed the lives of over a hundred people. Some of them were killed by police and had bullets in their bodies. The level of resentment from the protesters spared no properties. The sadness in Nairobi was illustrated by a signpost at roundabouts from Nyayo house which hung there for all of August 2005.

The whole of Juba and Rumbek where strategic towns and the late Garang had made several speeches supported his deputy Gen Salva Kiir Mayardit. Rumbek was also used as a reconciliation centre in case the movement had important issues to address. It was acting as a defacto capital for the SPLM/A.

With such shocking news, the SPLM/A which was relying on it leader experience, knowledge and skills had challenges of implementing peace ahead of them. Assurances were made by his deputy that the vehicle was irreversible and that he would take us to the promised land.

As he promised, the vehicle was not reversed and it was driven through and was witnessed by all of us to our independence from the north. President Salva Kiir Mayardit was the only living testimony after the death of Dr. John Garang de Mabior. He had survived his own thin and thickness.

The month of July was an omen. Pastor Mary Abuk Atem who was in charge of our church in Karatina called for an urgent meeting and organising of mourning prayers. We converged in the church by 7.30am and most of the mourners walked slowly but they had it in the church.

The prayers began with the singing of sorrowful biblical songs. Then I was given a chance to read out the names of people who perished with Dr. John Garang de Mabior. He had in the helicopter strong men like Lt.Col Ali Mayen Majok, Lt. Col Amat Malual, 1st Lt. Deng Majok Kuany, 1st Lt. Oboki Obur Amaybek and 1st Lt. Mayen Deng Mabior. The pilot was Col. Peter Nyakairu, Co-Pilot Capt. Paul Kiyimba, flight Engineer Patrick Kiggundu, protocol officer Sam Andrew Bakowa and air hostess Lillian Kabaije providing services to VIP and probably the Vice President of Sudan and Southern Sudan President Dr. John Garang.

The crew were alleged by Uganda's President to have had both night and day experiences. They were crew members of Uganda's Presidential helicopter MI.172-AF 605. The pastor was trying to comfort mourners with biblical references on the loss of Moses on Mount Nebo. Having lost her voice as well, Pastor Mary was struggling with the word that Moses died having delivered his people from bondage of slavery.

Assassinations of leaders such as Patrice Lumumba of Congo and Samora Machel had one evil commonality; assassinations of strong African leaders. He died in plane crashed having returned from a meeting called in Mbala, Zambia to give pressure to Zairean Dictator Mobutus Sese Seko over his support to Angolan rebels. The meeting was on 19 October 1986 and by night Machel decided to go back to his country against the advice of the Zambian Security Minister. His aeroplane was reported to have had a refuelling stop in Lusaka and headed toward Maputo. The crew members were employees from the USSR.

They were Captain Yuri Viktorovich, Co-Pilot Igo Petrovich, flight engineer Vladimir Noroselor, navigator Oleg Nikoaevich and Anatoly Shulipov. It was reported that the pilot and co-pilot had experience in both day to night flights.

Unfortunately, his plane went down on an South African hillside at Mbuzini killing him and two other ministers and some government officials. Only nine out of 43 passengers on board the Tupoler Tu-134 jetliner survived.

As matter of the territorial principle of International Law, a country would meddle in a crash investigation unquestionably. It was South Africa and to the great disappointment to the supporters of late Machel, it was the pilot who took the blame under an always common verdict. Not everyone was convinced including the South African Apartheid era former foreign Affairs Minister Pik Botha who called for a reopening of investigation. People like little Louw for reasons known to him or having acted under the influence of South Africa truth commission was reported to have come forward exposing the darkness of white men's dirty works in Africa.

As a former employee of the Civil Cooperation Bureau, he confessed that he helped bring about the death of Machel. The outcome of the investigation was mysterious. In the commission report it was claimed that the pilots cursed people on the ground because the crew were wondering why the ground lights were not there. He could not see anything as the plane was traveling at night and the flight data was blank. Those issues might have been where Louw was trying to expose what had happened previously.

In Dinka society, there is an expression that the death of one's own brother is an opportunity of his brother to decorate his hair by starting to wear ostrich feather. Separate from its use for decoration and costumes, the ostrich feathers are used to take off the dust or particles from electronic and glass materials. Like the inheritance issues of a deceased's properties immediately after his death, what happens if an African leader dies is a power struggle leaving many dots unconnected in the investigation.

Louw was working with the South African Civil Cooperation Bureau (CCB) as an operative. The CCB was operating under the Ministry of Defence as a death squad during the Apartheid era.

Like those who still doubt the outcome of Abel Alier's investigation on Garang's death, the majority of the Mozambiqians who attended World Cup that I met in South Africa way back in 2010 still believed that the decoy radio navigator had brought down the plane. In Uganda, as well as South Sudan, many theories have remained unanswered on the cause of the crash.

It was few years later that I had the privilege of meeting the late Maj Gen Atem Aguany Atem, a man who I befriended before death robbed us of his experience and fighting ability.

As a highly disciplined soldier and officer, who defended Aswa to allow me to cross to Nimule in 1994 with many bullet wounds sustained in the battlefield, he was willing to talk to me about the fateful day.

Maj Gen Atem rarely discussed sensitive matters but having known me through my uncle who lived three hundred yards away, he ordered one of the youngest daughters in his compound to get us tea. He was willing to talk. He said they were playing cards at their designated headquarters in New Cush when they heard the plane hovering overhead near New Cush's line clip on the south east part. They did not take note that it was the plane which was carrying their leader. The plane went south east of New Cush towards the Zulia Mountains and turned back as if it had wanted to land in New Cush. Then they heard an explosion.

Having been told that their leader did not arrive at New Site and that the Garang plane had got lost; he said all of them got convinced that it was the one which was flown over the ranges of New Cush and began the search. It was already dark in the area known for thick savannah and short mountain ranges.

Now, they had to act by mobilising enough soldiers in a hope to find their boss alive. They had to divide themselves into companies. Officers such as Chol Biar Nganga and Gatwich were dispatched to look for the wreckage south west of New Cush. Hours later they located the body. He hinted that the plane seemed to have been hit from the nose and the first persons to have been propelled by the force of the crash might have been the pilots falling harder on their heads. The pilots seemed to have fallen instantly before the plane landed at the tail causing a huge explosion from the fuselage.

He said they had reason to believe that the plane was brought down with an electronic remote control. He presumed that someone somewhere might have disabled the instrumental landing system and distance measuring equipment using sophisticated devices. Atem was sipping his black tea while I was waiting for the beautiful girl to stir my tea. He continued that the Kenyan and Ugandan authorities had also been searching. They could not find the wreckage and had to be informed by the SPLA that the wreckage had been located. The cockpit voice recorder recovered later indicated no forms of attempt by the pilot, co-pilot or navigator controller to counteract the situation. Their mouths were muted by unknown aliens as there was no communication from the pilots and ground controllers. Atem had it that external or internal actors might have disturbed the plane's control system.

He said having secured the site; they decided to keep the information secret until the top leadership of the movement made a decision on leadership succession. The bodies were taken to the SPLA headquarter at New Cush. Everyone was devastated but they had to accept that the Almighty God had taken the life of their leader. He said for twenty one years the SPLA men and files provided the most sophisticated close protection to the SPLA top commanders including its leader. One week after Garang's death, on 9 August 2005, one of Ugandan parliamentarians asserted that Uganda had violated its Civil Aviation Regulations for allowing the plane to take off in bad weather. He went by the theory of what many officials on both sides were gearing toward. And to go by that theory, he concluded that the regulation stated that no rotor aircraft (helicopters) should be allowed to take off after five pm for any destination lasting more than one hour.

CHAPTER 15

Karatina Church Emotional Experience

Inside our church in Karatina, the situation was becoming emotional. So Mary and I decided to ask people to disperse. I was the acting chairman of the Council of Church elders as my chair Akoi had gone to Nakuru for his studies. I concluded that Africa is a continent full of both external and internal conspiracies and a plane which left Entebbe at 5.02pm heading for new site would have arrived safely had it not been because of the likelihood of foul play.

Some members who had braved sorrow and grief had that those places had never any serious fog like the ones found in Russia and Europe during the winter time. Sometimes, the New Cush area temperatures were not very cold though one could at a time see thin like fog on the hill.

Ajak Manyok Ajak and I had lived in those areas before Ajak later became a member of the Joint Integrated Unit in 2005. We used to move between Natinga, New Cush and Lotuki many times either in the day or night nearly three decades ago. The Zulia peak is probably about 2,184m and slopes down toward Nefw Site at a lower 900m.

The independent research conducted by Atem G Dut discovered many discrepancies including the change of the initial route expected to have been used by the Pilot to a longer route. It was corroborated by the NASA meteorological report that on 30 July, 2005; Lotuki ranges were moderately hot for much of the day. It was about 28 C with scattered clouds and visibility at 30km from six pm to seven pm. There was enough visibility for the presidential helicopter reported by Ugandan authorities to have been modified with maximum versatility and to fly at night. Despite all those capabilities being repeated many times by Ugandan authorities, the helicopter could not make it to New Site.

In most of the conspiracies in Africa, it is African leaders who were willing to liberate or alleviate Africans from the overpowering poverty who are assassinated. As a soldier and a leader, Dr. John Garang was a believer of agricultural revolution on the continent and his speech during the Comprehensive Peace Agreement ceremony in Nairobi on 9 January 2005 might have sent a wave of fear to some African leaders who want to have a monopoly on continent control and the unchallenged looting of its resources.

Garang's concept of Agricultural revolution in Sudan, which could change the life of his people, was gaining international recognition. He could talk about it even when he addressed his soldiers who had just liberated Pochalla. His reasoning was based on the notion that strong food security is capable of supporting a weak government. He once had said that this would discourage past experiences in Sudan where people would take government positions to discriminatorily enrich themselves. He said that stronger people build strong governments. His philosophy was a challenge to the common definition of most of African rally intruders who produce weak governments which later become the peoples' burdens.

He was in one way a threat to Arabs. In *Emma's war: betrayal and death in the Sudan*, by Deborah Scroggins, the book had caught the Sudan Islamist government having not only been after Garang's life, but also after every living Southern in the oilfields in the Upper Nile. The book was given to me by Ustah Wach Duot Wach. Wach was an officer of the SPLA and fought on many frontlines before he became a Red Army (Jesh El Amer) teacher in one of the schools ran by UN in Kakuam Refugee camp. The book has detailed every account of Sudan's divide and rule dirty and bloody politics of the 1990s.

To understand it fully, I read it intensively for two months. Then I found a frightening assertion from one of the Sudan Ministers had that Garang's idea of a multi-ethnic new Sudan of equitable basis must not be allowed and that it was better that he was assassinated. In peace talks in Kenya, there were extreme Islamists who were not happy with him.

In Sudan's Civil War

In most of the assassinations in Africa continent, like that of Sankare and Muhammad Gadafi of Libya, it was always the Africans themselves who effectively executed the manmade evil act of terrorism and foul play. By the end of the day one would find Africans fighting themselves with brand new Chinese, United States, and former Soviet Union or United Kingdom made weapons. In 2007, two years after Garang's death and having left for Australia, I was invited to attend an UN Conference on the Gold Coast. With me was Sarina, a beautiful young lady studying a double degree in law and international relations.

The conference was organised by Bond University's United Nations club to deliberate on the impacts of small arms and negative roles of multinational companies in the developing countries. I was appointed to represent Sudan. My friend who was a master degree student, Benedict from Tanzania was asked to represent his country. As a representative of my country of origin, I did intensive research and by the time of my presentation I had in records that millions of small arms and light weapons worth billions of dollars continued to be sold to the Horn of Africa every year. The main players in this bloody game were the United States, China and United Kingdom and Russia.

Figure 15 during Bond University UN Club Conference as representative of Sudan, 2007

Sometime these superpowers would clash over dominance of the markets. When the diamond rich Congo gave birth to ungoverned territories, the powerful cartel would take over and start to control the direction of transnational flows of weapons. To ensure their relevance in the international arena, the speeches and statements of those who are purely Africans and seen as anti-west become the source of income for international and analysts.

Leaders like Dr. John Garang in the rebel movement might have provided fertile grounds for those who wanted to trade off his life for anything including money.

The plane Dr. Garang used was made in Russia and proudly praised as the Ugandan President Museveni's helicopter. While watching Museveni's speech from the television during the funeral of our leader and his bodyguards, Museveni was somehow hinting foul play, but he was not making conclusive the probable certainty of external factors. Instead of attending the funeral in Juba like most of leaders in the region, he did it from Yei. His minister of information reported that he was too devastated to attend having lost a long term ally and friend. The family members and the leadership of the movement of the late Garang were stronger than the Uganda's officials? Mama Rebecca who had to console the mourners to be strong and that peace would still be implemented in the absence of her husband was a lioness who encouraged everyone during our dark hours.

Now, the entirety of the crash was left to the investigators. Everyone around the world was watching the investigators unearth the facts. Sudan which was using an iron fist to clear protestors was panicking. Their fear was the likelihood of the country going back to war. Darfur which had seen low intensity conflict in 1991 before Daw who was sent by SPLA/M leadership was murdered was already two years into an armed conflict against the Sudan Islamist government. So they had their own reasons to conclude the investigation committee quickly.

Uganda had to assist the committee and international experts on the crash investigation were drawn from Russia and the west. The crash site was inspected and finally the black box was taken to Russia. Everyone was waiting for the outcome and every day I was on the internet café in Nairobi looking to read something about the outcome of the crash investigation. Several months later, the former Vice President and President for regional government in the south Abel Alier made the result public in a precise summarised form. The verdict was that it was pilot error and bad weather and that for the time henceforth the file remains closed.

Like most of investigations hidden under a fog of conspiracies, the dead pilots took the blame. Other newspapers, which I read, which gave different accounts of events were censored. A statement by one of the members of the investigation committee former Interior Minister on the possibility of foul play during the flight nearly caused him trouble.

Africa's politics is that if one wants people to destroy evidence it is the brother of the victim that is allowed to do the talking.

In the investigation committee, we had our own sons some from irreconcilable careers with the investigation at hand. Some for instance had experience and knowledge in law and military science with likely zero knowledge on plane crashes. While reading the result of the finding, I was like if I were the one sitting in the Russian centre of crash investigation what else would I have known other than to take my trip allowances and relied on international experts? Before he died, Dr. Garang had to survive many attempts on his life.

The United States' Former Envoy to Sudan at the time was already ahead of the committee in determining the outcome of the investigation. He was heard blaming bad weather, poor visibility and pilot errors. In the trial of the murder of a primary student trial in Brisbane Magistrates that I attended as a trainee, the jurors empanelled for that purpose were cautioned not to rely on prejudice either from the mainstream media or any other statements which have a potential to impair their judgment. The outcome of investigation was being pulled to a direction to suit various interests.

CHAPTER 16

Resettlement to Australia

DENG GARANG BUL was in form two with us before he had left for Australia. He promised to take me to Australia and that he would send me the humanitarian visa form. He knew that I had registered as a refugee in Northern Kenya. And that I had the intention to relocate to a third country as was my right in the UN Convention on the Status of Refugees.

By February, 2005, before the form application and related documents arrived in Nairobi, there was other news of our results. I was told that only Dhuol Mac and Paul got B-. Before the results were published, I had a dream similar to what I had when my Primary School National Examination results were about to be released. In the middle of the night, I had a dream. I was shown two grades by someone who looked like one of my teachers. He was not clear on what he was telling me but he had pointed to higher grades before moving his middle finger to C+.

I quickly took money from aunty Clementina and rushed to Arnesen. As revealed in my dream, my grade was C+. People who were informed about the result included my friends, relatives and family, and siblings.

Like my classmates, they congratulated me and encouraged me to move on. My sponsor and brother Mayen immediately called from the USA to find out the result. I told him I was not in good mood to say anything with him having spent so much money for my grade to come out like that. His replied was even more impressive, that if wishes were horses, beggars would ride. He concluded that it not all what we truly long for would be granted to us.

Having passed through thin and thick at a delicate age and missing school for several years, those grades were achieved through hard work and dedication. Among us there were Kenyan students who started from preschool but their grades were so disappointing to the school and parents.

I accepted it as a gift from God and appreciated what was at hand. I then said many people I left in the war zones had been denied such opportunities even to attend primary school. I thanked everyone who had stood by me from parents, brothers, to uncles and friends.

I started with my classmates who used to encourage me in time of difficulties. In the house at Karatina waiting for me was the special humanitarian visa application forms. Having collected my result slip, I rushed back to Nyeri.

I filled in the special humanitarian (offshore) subclass permanent residence visa. I had all the reasons from persecution in Sudan and the experience of the horrors of the civil wars. I almost a wrote a novel. The story was very long and I had to use extra papers as an attachment.

The application was lodged through Australia Embassy in Nairobi around February 2005 by Dau Garang Aleer Abit on my behalf. Attached on the top of the application were two passport sized photos. The photos were so blurred that I was worried that if an assessor with prejudice against Africans came across it then they would conclude that a grade one dropout and drunkard was about to make his way onto Australia's soil. The land's distant wars and conflicts were wrongly reported by all the Western Media as embedded in people's blood.

My eyes were not clearly visible and looked like that of the rotten giant Nile Perch. The clothes was also second hand from America with the 'raised fist' symbol of black power. The application reached Sydney by air. With nothing to do I continued exploring the big cities of Nakura, Nairobi and Nyeri.

One day as I was crossing from Githeria bus station near India temple, I came across two street boys not only robbing their victim, but urinating on him so that the man had to hand over all his coins to the Nairobian town gangsters. He freed himself and escaped. Mayom Atem could not believe how people lived in major cities handled such unfriendliness.

After two months, I received a file number and subsequently I had a call to attend an interview at the Australian Embassy in Nairobi. I neatly dressed up with blue trousers and a grey shirt.

With all my supporting documents with me, I had to make sure I went with the Kakuma school teacher Thiong Akech. I politely requested him to act as my interpreter. I told him that my English is not at the standard of the Australian Migration Agent and requires someone who speaks with an American accent.

He agreed, however, he said that Nelson Mandela had once said to those who loves anything foreign that if one speaks the language understood by the rest, the words go to the brain and, if it is one's own language it goes to the heart. I wanted Thiong to ensure what the interviewer was saying goes to my heart. Sometimes I had to reply in English so that he did not think that Australia was being packed with semi illiterates.

By seven am, I was by the gate of the Embassy. I met a few applicants who too were waiting for the gate to be opened. The gate was opened and I was allowed to sit to wait for my turn. The interviewer introduced himself and was very nice and an angel of potential identification. He knew that there was hidden potential in me. He said that I passed the interview and asked that I grab form 80 from the Embassy counter. Those who failed interviews were banging their heads outside and crying.

Some of them were recalling the harsh life of being a refugee and stateless and to make it worse in foreign land. Duot Atem, who failed his interview had his case become a full month community dedication to God. At our local church, prayers were being organised by our Pastors.

Then I had my medical checkup. My blood was tested and everything came out negative. A physical examination was done which included everything such as checking my genitals and dental inspections. The examiners did not know that they were committing terrible mistakes here and there. Inside the examination was a beautiful woman, a doctor who had studied everything about human anatomy. She ordered that I undressed down to my underwear. She beat both joints and listened to my heart beats with a stethoscope. Finally, she said that I was fit for purpose and should put on my clothes and go. Then I left and took a bus to Yaya Centre where Dr. Dau Aleer Abit had rented a house for his family.

Unbelievable to us was the miracle which worked out on Duot's application. Instead of the rejection which he expected, he received his approval together with me on the same date of 20 August 2005. All of us were delighted. I was like there is likelihood that the miracles had worked either at the Embassy or Sydney. Natural intervention and some invisible good guys somewhere might have reviewed his interview in his favour.

Overexcited, he decided to rush to Kakuma for a celebration with his relatives upon the news. To organise the celebration, he had to borrow money.

He bought a goat back charged against his empty future account in Australia. He was optimistic as he was leaving life in the camp. We heard the story of a man who upon boarding a plane at Kakuma town's dirty airfield, cursed the soup of lentils saying never again would he eat the food eaten in refugee camps. The ideas many had imagined about life in the west had been that of islands lived on by angels with no issues: the presumed paradise in heaven.

Duot's celebration was well attended by respected community elders being the son of the famous Ayual local singer, Atem Bul. As my nephew, he was willing to share with me his good time in the camp.

I decided not to go to Kakuma because I ordered my elder brother and the rest of family members for Nairobi's celebration. In the camp, it was a custom that anyone going to a foreign land like the first batch of lost boys and girls organised a celebration so the community would have time to advise them so that they do not forget their people and culture while in the west.

As usual, Mayen approached maternal Uncle Malek Arok Deng to borrow money. As a result, Malek approached Abraham Awai Piok and an one way ticket was sent to me. The flight details were that I would leave Nairobi on 4 October via Dubai, Singapore and finally Brisbane in Queensland, Australia.

By ten pm, I was driven by my late uncle Chol Aleer Abit in the company of my uncle Aboui. I arrived at the Jomo Kenyatta International Airport. Then the Kenyan migration officials took the travel document and stamped and cleared it for departure. The document had the Australian permanent residence visa on it. I was walked into a plane by the crew member.

By about twelve am we took off and arrived in Dubai early in the morning on 5 October. We had a rest inside the very neat and clean airport and by ten am, I had boarded an Emirates flight for Singapore. As usual, the security screened my bag. After an hour we were called into the plane at about eight am and on 6 October I arrived safely in Australia. I was cleared by the migration officials as I got welcomed to Australia.

Outside the airport was Pastor Deng Abot Riak who was working with Anglicare, a church organisation which helps new immigrants. Besides him was Malek Arok, Mawut Mabior and two other gentlemen.

I was driven through Brisbane's beautiful interconnected highways.

The traffic lights were doing a human's job busy flashing to motorists on which one should stop and go first.

While we were using torches to assist SPLA drivers pass through busy roads at nights, there were other places where the lights automatically opens to tell motorists what to do.

Then we arrived at Brisbane city centre. It had amazing and beautiful skyscrapers and modern tall building. Near Queen Mall Street, I saw a blind old man being guarded by a giant dog. It was unimaginable to me that a human being had trained a dog not for hunting but for helping a vulnerable friend. I asked Deng to stop as I had a close look at what was happening.

There was nicely decorated harness wrapped around the dog. On the back of the harness was a leash or what seemed like part of a chain with a loop for a handle. The leash was used by blind person to handle the dog. When the traffic light was red, the dog would move closer to the pressing button and push it. Keenly watching, the dog pulled the person upon hearing the sound and green light and they crossed. I was stunned.

We arrived at Akech's residence in Brisbane for a rest. I was warmly welcomed. Everyone including his nephew Jacob Chol concentrated on the welfare of people back in Africa, Kenya and Sudan. After several tiresome flights, I had made it to the new continent.

It was 2001 while I was in Kakuma under a tree between our compound and the Ayual community library that Dit Nyuon Abui came to me with a small map of Australia. It was sent to him by a pen friend. The map had states, kangaroos and koalaa, oceans and islands. I looked through the map and saw a picture of giant kangaroo with it young ones in the pouch. Then Dit and I were joined by three people older than us. These people confidently said that the kangaroo was a powerful animal which could jump and fly up before landing several hours later. They concluded without hesitation that it is not like the lion, tiger and hyena, they would say that if the kangaroo wants to kill a human, it would put the person in the pouch alive and ingest that person before swallowing it. Upon hearing that, there was already well informed fear on how dangerous the animal was.

In Brisbane, the delicious variety of Australian foods included KFC chicken. At four pm, we left for Toowoomba, a small remote town east of Brisbane. It is the largest inland city in Queensland on top of the western slopes of the Great Dividing Range. Known as a centre of commerce, industry and education, Toowoomba is the regional capital of the Darling

Downs. I met a lot of people like Machar Piok Keer, Atong Ayii Jok, Aluou Ajak Atem, and Rev. James Kuer Ajak among many people from the Greater Bor community.

Finally, I was taken to my uncle's place. Deng Garang, my sponsor, was at university doing his Bachelor's degree. He knew I was coming and therefore had to arrive at eight pm. He was followed by some of my old friends who converged at my uncle's place. The distinctions between the two continents continued to midnight. While it was midnight in Australia, Africa was way behind for hours. By four am, I was asleep and I woke up at nine am.

There was already breakfast on the table waiting for me. I had two cups of porridge. Then Rev. Deng Abot and my uncle Malek asked that I go to Centrelink. Rev. Deng was the driver and we arrived at Centre-link complex. My details were made with fortnight settling sort of social benefit and the Australian Federal Government's statutory payment to a newcomer until one finds a job.

Thereafter, I was driven to the bank for my details which were recorded and I was provided with a bank account for the first time. I thought it came with money as well and that I would work and save the money in that account. But it was not the west that we were expecting while I was still in Kenya: the other planet where one assumes that everything even the account would be full of dollars. It was an empty account and I had to stock it. The difference was that the west had an abundance of opportunities and prolonged stability and steady democracy. The hard work to prosper in life is the same.

Then we left home, had our lunch and dinner in the evening. There was a fridge inside the house where dinner leftovers were kept inside for anyone waking up at midnight to fill his or her stomach. My stomach had to get used slowly on how to not get up at night to fulfil the desire of body. My weight began to increase. I was taken to the Technical and Further education (TAFE) college after a week. I presented my primary and secondary school certificates to the teacher in charge. TAFE was offering flexible programs which if one meets them enables the students' entry to university.

There were English courses for migrants at TAFE though the one I attended was like a money making machinery for those who got employed over there. My age was assessed and I was informed that I am an adult who had to go to high school and should continue refreshing my English for one semester. I had missed education and it was unimaginable how thirsty I was for knowledge. So I accepted the offer without hesitation.

Our class had both mothers and grandmothers. I was with senior mother Achol Aquin whose brilliant son had graduated in Information Technology from the University of Southern Queensland. His older son was reported to have been murdered by Arabs in the '70s during the Southern Sudanese strikes against Arab dominance and the ruthless spread of Islam in the Southern Region.

The class was set up in such way that there was no difference between those who had not gone to school and form four leavers (year 12). Also, being foreign does not mean that one is not smart.

I was told by migrants who had been there for over six years that it was akin to the story of one of the students who studied a nursing degree at University of Khartoum. He was said to have spent twelve (12) years studying and so the dean of the faculty had mocked him by asking whether he was studying the elephant teeth.

At TAFE, we were taught the sounds of vowel letters. It was how to pronounce word like a white academic not the type of slang I found in the Toowoomba shopping centre.

This was a joke. There is a saying that one cannot teach old dog new tricks, or in Dinka, they would say one cannot dehorn a bull when they had already reached the climax of its maturity. I protested and submitted my certificates for the second time to the head teacher. All those requests were brushed aside and nothing was done. I was required to attend or payment would be cut at Centrelink.

I was still being accommodated and ate free at Uncle House but also a looking for independence, I left TAFE for the university pre entry preparation courses. I asked Garang to provide me with an application form. By seven thirty pm, Garang Chut Deng came with the form. He gave me advice on how best to fill it. I filled it overnight and in the morning, I rushed to a mall nearby to secure the authentication of my certificate by a Justice of the Peace. By eleven am, I was already at the University of Southern Queensland. After one week, I started the bridging course having secured an offer. Then I started with mathematics among others.

I was also waiting for the outcome of my 27 November 2005 application to five universities for a Bachelors Degree. In one application on the basis of preference, I had Bachelor of Law, Sciences, Criminology and finally International Relations. Besides, there was family financial pressure coming from those I had left behind.

So in December I went to the Toowoomba Greyhound bus station. I bought a $75 ticket to Dirranbandi.

I passed through the Lockyer valley adjacent to the Toowoomba hills. It is rich farmland between Toowoomba and Brisbane.

Then we proceeded to Gatton, Ipswich and finally Rome Street Station in Brisbane. We rested for few hours then travelled for about five hours before we arrived at the town of St. George about four hundred sixty seven kilometres from Toowoomba. At six thirty pm, we drove to our final destination. The area is home to the largest cotton plantations. I got a job at a cotton planation and then made a bit of money.

The payment was made in the field upon the arrival of supervisor. It was money without bank charges and taxes. Before the company car arrived, we would make sure we woke up at three am. In the evening we would proceed to St. George Town for refreshment. The temperature was averaging 33.7 and 35.4°C being summer time in December to February. I sent the first payment to people back in Africa. The problem is that most of the Africans in the diaspora face are paying rents, electricity and feeding while caring for others in Africa.

By five thirty am, we would have already started and would work until three pm sometimes to four pm. We would make sure we had our ice water and food ready in the containers. With some reckless drivers, we would find kangaroos crashed by cars either in the middle of the road or on the sides. Some were blinded by the high beams and ended up standing and confused in the middle of the road.

Then the old story in Kakuma of kangaroo jumping several kilometres after having snatched humans was proved to be false. In the surrounding forests, were fast moving wild goats spotted with our car beams. I asked whose goats were those? The answer was that they were food for the dingo, a sort of wild dog.

On 14 January 2006, I received a phone call from Bond University. I was called by admission officials. Among them was Nosa Esiet. He is Nigerian scholar, who studied and had lived in Australia for years. He said based on the assessment of my high school certificate, which did not include Kiswahili during the admission board meeting, I got a provisional offer to study law. In addition, he concluded that I should make an acceptance with immediate effect.

I immediately informed my close friends of Akol Maluk and Chol Akoi Jurkuc and other colleagues of my intention to wind up the manual work at Dirranbandi and Goodiwindi. Akol like my uncle Malek and Chol were very encouraging.

Since I had arrived in Australia, my uncle had continuously advised that I should continue with my dream of becoming a lawyer.

Since my dream was deeply rooted at the time of my father before he passed on, I replied to an offer without indecision and then took off for Toowoomba. I spent 15 January 2006 in Toowoomba and had to arrive at Bond on Monday 16 January 2006. I presented my admission letter to the school and the completion of my admission immediately began. Then it was already four pm and I had no idea of where to sleep that night. Also, I had little money which was not enough to book hostel for a week and buy stationery. It was a rushed decision fuelled by excitement and my prolonged desire for quality education.

Back in Africa, only the sons of corrupted politicians, senior military officers and rich business people have access to high quality goods and services from abroad. Those who send their children with clean money are few.

And those whose fathers had passed on like me when they were even peasants rarely see anything of western countries except for the large cargo planes which leave thin lines of smoke in the sky. Now equal to any son and daughter of those who had their majority of constituents poor in places like diamond and minerals rich Congo, I had not even the slightest trouble in my heart and was ready for classes. I did not know that Bond was not Arnesen and that the era of shining shoes with Kiwi shoe polish was over. The Australian environment was clean because most areas were tarmac.

Then Nosa came. He convinced the school security guards and I was allowed to sleep in one of the rooms at the hostel. I was worried that if at sunrise, I would receive a crazy bill. I knew I have good people like Nosa and Prof Tina Hunter from Norway supporting me so why should I worry?

That was the first time I slept at a hostel reserved for the children of rich people. I was given my student ID and government financial assistance form. The next day Nosa came and took me through the processes. Then I went to the student admissions office near the University's club. I was furnished with the timetable for semester one. Australia's legal system, Information technology, Contract and Tort law were my subjects on the timetable.

I was given a form for student loans. I filled it out and after a few hours, my student identification card was ready for collection.

I was so excited that what I had been looking for years had arrived. I was proud of myself for having made it to one of best law schools on the rich Gold Coast.

About ten minutes' drive away is the Pacific Ocean. It is about 9.5 kilometres from Bond. Its beach attracts hundreds of thousands of tourists worldwide. The Q1 tower the tallest residential building in the southern hemisphere stands at 1058 feet. I would take a lift to access the Q1 tower with its observation deck facing the Pacific Ocean. The Ocean and Gold Coast views would be clearly visible from the telescopes. Adjacent to it were Surfers Paradise, and the Oasis and Pacific shopping centres.

Figure 16 from right Deng Malual, Dhieu Magot, Mike Majok, Ajak Atem, Awai Piok, Thieu Atem, Ajak Deng and I on the Surfers Paradise, Gold Coast beach, 2011

At the end of Bond University Arc is Lake Orr. It is a man-made lake connected to the Ocean. A drunk student throwing themselves in for a swim has their fate determined by stray sharks.

The University during the summer attracts both domestic and international students. Some come because of its tourist like environment. We rarely missed exchanging ideas and intermingling with those from Europe, Japan, USA, UK, West Africa, Latin America and even Middle East every semester. The brilliant students across Australia such as my best friend and once classmates Mathew and Maria would both get onto the Dean's list and enjoy the warm weather at Robina town.

By 3pm, Nosa came directly to me and proposed that we look for shared accommodation around the University.

So we went from house to the house. We were like police officers on a search warrant mission. If the school had not already created a conducive environment with, we would have been looked down upon with all sort of discriminatory common stereotypes against black people. Nosa, short and huge going around with another tall and skinny black guy would have attracted police attention if it was another country or suburbs. Undoubtedly, the home owners around were also looking for ways to make money. They were also benefiting from renting to both local and international students. We found one available bedroom.

I introduced myself before my African greeting of senior persons of her age as mama almost ruined the contract. I immediately apologized. Nosa had to intervene by saying she was talking with a fresh entrant from Africa. She smiled and added there would be more culture shocks ahead for me. Then the oral agreement with the lessor was that I should pay $600 a month including food.

I took the room key, put my bag and did the setup of my reading table. Over those ten years, I have been in great debt to Nosa. He was doing everything possible like my elder brother and African scholar.

At seven pm, I went to school library. Every day, I would read until nine pm before the law library was closed and I would move to the underground library. At midnight, I would still be preparing to retreat to my queen bed. On Tuesday 17, 2006, I started my first class on the Australian legal system. I came in five minutes late and I almost got blinded. I saw a giant screen which covered almost a quarter of the section of the wall with lecture slides by Prof John Wade. It was something I have never seen before since most of our writing in both primary and secondary schools were done in exercise books. Teaching was conducted using chalks and blackboards. All the lectures at Bond were streamlined. The next day, we attended tutorial sessions.

It was a total shock. Waiting for me at home was Australian food. I started on some serious reading, slept at three am and woke up at six am. I was losing weight. That was not a good sign because in our village being fat was associated with wealth and well fed sons or daughters of dignified families. At school, I met new friends.

The teachers were also making sure students' confusion was minimised. It was not funny that students were moving and running randomly from one building to the other looking for lecture halls. I enrolled in mastering public speaking for leaders with Professors John and Tina. It was fantastic. In his class, each student was given a chance to address his or her classmates and lecturers on chosen topics. It was aimed at boosting students' confidence and the shy ones would stand speechless in class. We were taught maxims of public speaking and about the role of adrenaline when the speaker is about to speak to a crowd. Then I like many of my new friends immediately adapted to university life. I did my first semester under extreme pressure. I was struggling. It seemed like I had jumped too many steps to law school.

However, I had Prof Tina Hunter and Joseph Crowley beside me. I remembered her advising that I should seek tutorial assistance and I did. Against all odds, things began to work very well. And those I had left behind at the plantation began to have confidence that law school was not only for A students but for those who worked harder.

In mid-February I was introduced by Nosa to the Dean of Law Prof. Duncan Bentley who had lived and taught in Apartheid South Africa. He is a devoted Christian. He is a kind hearted and committed person. Every Sunday, Nosa and his wife Pearl would pick me up at my shared accommodation to attend church services at the Mudgeeraba Uniting Church.

I narrated my life story to him. Immediately, he encouraged me to apply for a scholarship since Bond University was one of the most expensive universities in Queensland.

I did and my school fee was slashed by half. It was a scholarship which was later continued by Law Faculty's Dean Prof Geraldine Mackenzie for the Post Graduate degree of Legal Practice for Australia's qualifying lawyers. Then came the second semester and I was offered a part time job at the Bond University restaurant replacing Donald who had just graduated from law school and left. To cope with the time pressure, I would attend my lectures from morning to four pm. Then I would start work from five pm to ten thirty pm.

I worked there for a year. I was able to save some money to support myself and relatives in Africa. I took care of my cousin Ayual Kuany Alaak who despite the best treatment provided to him at the Jomo Kenyatta hospital, could not survive.

I was promoted to second year; however, I decided to defer my studies for a later date. But first, I had to consult with Nosa and my administrative and constitutional law professor Keyzer. I came to know the professor when he taught me two subjects. I told him that I needed his advice and that I wanted to defer law degree for one year.

Figure 17 Thon Abuoi Thon and his wife Alek Bior Atem ak Mama Anok inside his house, Perth, 2014

Back in Africa, there was a beautiful girl I had fallen in love with in Nairobi. I used to talk to her on the phone in the middle of the night. Sometimes her sweet voice disturbed my sleep. Her voice had not been that sweet when I was a Nairobi hustler pulling dust to their house at Githeria 43. Girls' pretences were so common that in those days a few friends of mine would find their spouses being picked up at the airport by boyfriends. I need to prove that I had made it to the land of opportunity. The cost of printing lecture handouts was the same as how much it cost to call her.

But back in the day, she would say I had not been her type because one of the lost boys had made a first booking. But this time around her usual rude voice begun to disappear and there was no more ambiguity.

She had suggested that if I was serious, I should make a formal marriage proposal through her parents. This meant paying thousands of dollars in kind and I googled for available jobs across Australia. Then I phoned my cousin Thon Abuoi Thon, and there was a speculation that to be rich, one must work in the mining sector in Perth in Western Australia. That was where my thoughts were directed.

The professor responded positively. He said that in his lifetime as a senior teaching professor, some of the students who left school rarely came back for their studies and on no grounds should I make that mistake. He was telling the truth since two students I had enrolled with at the same time had vanished.

I found one loitering on the streets of Surfers Paradise in the summer of 2008. I chatted with her for five minutes. She was heavily pregnant.

As usual, Nosa loved Africa and one day he organised a dinner at his house. This one was attended by people from Africa such as Nigeria, Ghana and South Africa. While we were eating, a certain lady asked why I do not have a white girlfriend. She said she has been seeing me without a girl for some time. She was joined by the Ghanaian engineer who had broken up with his white girlfriend and had split everything with her.

I laughed before I said the pressure I am under is not like anything I had been through before. These kids have never seen anything bad. They have never been bitten by flies let alone mosquitoes and scorpions.

I was also thinking of Professor's advice that I should study hard and complete my degree and the rest of goodies shall follow. The school had assigned students academic advisors. They were great people who encouraged hundreds of students to overcome enormous challenges from family to personal life.

Prof. Keyzer was reviewing his book on Australian Constitutional Law and working on behalf of the Federal government on funded research regarding the Australia's Interdiction Policy of boat migrants on the High Seas and its harmony with International laws of High Seas by the time I approached him with my personal issues. He immediately called his secretary, who was a student with us in one of the subjects. She was asked to take down my details.

The next day I was called to sign my part time contract as a research assistant at Bond University. And was told that I would be reporting to him. As a researcher, I would explore anything from all the conventions and treaties on refugees, stateless persons, their rights and protection. We worked together and I learned a lot from the Professor until we ran out of funding in 2009.

We applied for grants from Queensland Government on capacity building for new migrants already on Australia soil and crimes related to refugees. The grant fall under the Criminal Proceeds Confiscation Act 2002. The Queensland government decided that any money or properties gained from criminal activities should be channelled towards the advancement of quality of life through research or local organizations which work to empower vulnerable groups. While at Bond, I met many caring people like Senior Teaching Fellow Joseph Crowley who taught me in more than five law subjects including advocacy. He is a High Court Barrister specialising in criminal prosecutions and international criminal trials and supervised me on two independent research papers on war crimes against women and genocide in Sudan's Darfur region.

By third semester of 2008, Joseph was our coach on proposed International Criminal Court Trial Competition in The Hague. He set up a team of bright students and great public speakers. Then he put them through training. He divided them to Victim and defence Counsels and prosecutors.

Figure 18 with Kate Michelle and ICC Bond Team Competition 13 Feb 2009

In our first year in the law school, we acquired skills and knowledge of court proceedings, which used to include appearing before the judge at magistrate courts to represent a case developed by the lecturers with either the Brisbane or Gold Coast magistrate courts. At Bond, I met Prof

Laurence Boulle who later became the Dean of Law, a mentor and friend. He is an international expert on Mediations and Dispute Resolution and has written intensively on law and globalisation. I also had the support of Louise Partson, who was an Assistant Professor and my tutor in most of the land and properties law subjects.

In February, 2009, Bond was invited to participate in The International Criminal Court (ICC) in The Hague.

I was in the team participating as an observer. I joined the team when practices were almost completed and had missed out on a crucial part of the training and practices.

To fly out of Australia, I need an Australian passport. I had to book citizenship examinations which were done online and tested candidates' peoples' history, values, democracy, responsibilities and privileges as citizen. I only missed one point scoring 19/20. This test was better and not like when I sat for the Officers Defence entry course to join the Australian Navy when the mathematics of angles and degrees failed me terribly. This was before I moved to the fitness test. In our days in the bush, being a soldier was based on fitness and how tanks crushed their victims was not important. What was important was the defeat of the enemy.

In the citizenship test, there were a few questions about the Aboriginal and Torres Strait Islander peoples of Australia. These were the original peoples of the land before the white settlers and convicts invaded and took their lands. Those people suffering remains beyond any imagination and Australia is still a colony, though currently self-governed as there has been no treaty signed.

For over two centuries until a referendum in 1967 they were not considered human and were classed as fauna and their children were stolen from them and they were murdered and moved off their land and forcibly converted and imprisoned. These has killed off many of the different languages and cultures of certain tribes and has meant that many descendants who were stolen away as small children and forced to work as servants do not know who their relatives are or what tribes they came from.

After the referendum they were finally allowed equal rights on paper and could vote but for many this does not actually happen in practice and they face constant racism every day and many organisations and services are set up in ways that prevent them accessing them. They are required to learn English in schools and are punished for knowing their own languages instead and many are locked up for extremely small and minor offences due to massive racism in the police and justice system.

Nothing much is discussed about them in many books. My lonely citizenship ceremony was conducted and the next day I was issued with a passport. On the 25 January 2009, our team under Coach Joseph Crowley boarded the British Airways flight travelling the whole night. We had one stop at Singapore's Changi International Airport for one hour before we braved our long flight of 21 hours and 50 minutes. We passed over Mountain Everest and the navigator showed the whole map of Europe where old cities and countries such as Warsaw and Sarajevo, Finland, United Kingdom were visible. The appearance of those cities and countries including Berlin revised my memory of the two major world wars and the partition of Africa, which I studied in history back in Kenya.

For instance, I saw nothing at Sarajevo other than what transpired after the assassination of Archduke Franz Ferdinand in 1914 which later ignited the First World War. In high school, we studied the invasion of Poland, Warsaw, Finland and Russia by the forces of Adolf Hitler and his allies.

I was not sitting idly inside the plane for I was busy recording and connecting the past events with my new adventure in Europe. Two months before, Joseph who convinced the school to provide me with finances was insisting I go to Europe which he said would be an incredible experience. And indeed it was fantastic.

By the morning of the next day we arrived at the London Heathrow International Airport before we crossed above the North Sea passing over Calais and finally to Amsterdam International Airport. The North Sea is located between the eight European countries of France, Germany, Netherlands, Great Britain, Scotland, Denmark, Norway, Belgium and France.

It was a cold season in Europe and the visibility was very poor. The fog covered the sky and airport. I do not know how the pilots managed it.

I had to put on heavy clothes provided to me by Joseph. We booked a hotel. Instead of Australian food, we were eating bread with soup every day at The Hague.

The next day, we took a tram to where the competition was taking place. There were students from Europe, Asia, Africa and America ready to show their wits.

The trial before the judges was on issues such as genocide, crimes against humanity and war crime in specified localities where two belligerents fought killing civilians and engaged in the pillage of properties.

There was targeted killing based on ethnicity. The competition took a few days and the result came out. It was a gigantic celebration as our team had beaten Yale University to become the World Champions.

The next day we took train to Amsterdam from its station at Den Haag via Leiden. We paid 13.60 each. It was about forty minutes before we arrived. We took a walk around Waterlooplei, foam, national monument and Hermitage museum. We learned a lot of historical facts before new realities such as what I saw at a sex shop revealed the conflicting realities between the past and the present of Holland.

We did some shopping and went back to our accommodation. The next day we went to the Permanent Court of Arbitration called interchanged after the International Court of Justice (ICJ), International criminal Court (ICC), and International Criminal Tribunal for former Yugoslavia (ICTY) complexes.

The ICJ is the international law administrative building in The Hague. At the gate, we were cleared by the security guards.

At ICTY, we were taken around and finally to the desk where proceedings against leaders of the Serbia- Bosnia war were taken place. These were boring proceedings which involved watching videos. After thirty minutes we left for ICC where we also found Congolese warlord Bosco Ntanganda answering charges of war crimes on the forced recruitment of child soldiers through his defence lawyer. He was later sentenced to thirty years seven months after we left The Hague. Upon arrival at ICJ, I saw visible signs of when the peace place building was constructed.

According to the employees in charge of the building, it was built in 1904 with the intention to encourage states in dispute to solve their disagreements without having to resort to armed conflict. The building was made of precious stones and each participating state had to bring in their best gifts following the series of Hague Conferences between 1889 and 1907. The inside shone as if it was made of gold and diamonds. Clearly visible was a vase from Russia standing firmly and shaped like a world cup donated by Nicholas Lancery II. It was made of what looked like green jasper mixed with ornaments. The decorations included lion masks and two headed eagles and the Romanov family coat of arms. I adored such a work of art. During our revolution, our first supply of weapons including military gadgets came from allies of the Soviet Union in Africa such as Libya and Ethiopia in the mid '80s.

At the entrance and inside, we were briefed on issues before the court such as Russia-Georgia and Sudan Abyei disputes. Abyei is home to the Dinka Ngok people which have become a centre of terrorisation from Islamist governments for almost a century. The dispute was between Sudan governments and Government of Autonomy Southern Sudan (SPLM) which was later decided in favour of Dinka Ngok though with redrawing of their original boundaries. At our dinner table, I was asked by Joseph whether I wanted to go with them to Rome, Italy. My reply was that I wanted to go back to Australia to catch up with my studies.

On 23 February, I had made it safely to Brisbane International Airport. At the exit, I was singled out by police.

I was then taken to where a racially motivated search was conducted. The sniffer dogs were brought in. They quickly ran through me and my bag. They found nothing. I was in shock. The rest of passengers were allowed to pass through. Then a female police officer put on gloves and forced her hands inside my bag going through my clothing. I had to challenge her intention and she said it was her job.

Then I replied that if her job at the Airport was to check male underpants, then she was doing a stressful job. Then the policeman standing next to her laughed in a mocking way. He came closer to me and said you know, there were many cases of illegal drugs trafficking in Holland. I told him that what they did was wrong and that I just came from representing Australia for you to racially profile me and that it was unfair. I grabbed my bag and left in great disappointment. Waiting for me outside the exit entrance was Alaak Deng Ajang.

He drove me directly to my brother Ayual's house. It had been just two years since he came to Australia with his wife Khot Ayiik Bul and two boys. After a week, I was joined by the team. Immediately, we commenced our classes.

The university administration organised a celebration. Ever since, Joseph has been on my side assisting me in all aspects. I spent about three months then we left for Africa with Tom and Hellen from Australia. We arrived in Nairobi and then proceeded to the tourist hotel around ten minutes' drive from Nairobi city centre. Tom and his friend decided to go to Botswana. I boarded a Jet link flight to Juba. I have only heard about Juba in the 1990s because of wars and massacres. It was my first time going there.

I was picked up at the airport by Uncle Driver waiting for me in his full military gear. He took me to the dirty old Juba centre. He said by the time they arrived there the city was so dirty that no one considered it a town. The Arabs who lived there since its inception knew that if the owners of the land came, they would leave or be forced to go.

So they deliberately decided not to develop the town. Some areas manned by Sudan Army were being controlled by the SPLA.

The Comprehensive Peace Agreement was that Sudan Army Forces must be deployed at Sudan-Southern Sudan borderline as it stood in 1956. There were only few soldiers from Sudan Army under the Joint Integrated Army.

I spent three days and moved to Bor by bus. I passed through the very places I had walked barefooted in the 1990s. This time I was on a bus called Nationdit "Big nation". Southern Sudan was about two years old in determining its political destiny through International Monitored Referendum. The referendum was meant to determine whether Sudan would split into two states.

I arrived in Bor where I met my brother Alaak and his family at Laudit.

I went to church packed with politicians including Jonglei State governor. I was given the floor. Then I made a few remarks which were purely political and explained the war crimes and genocide charges against Sudan President. I could say in Africa, if one wants to castrate his dog, the best place to do it would not be in his home as the dog would run away and not come back. The dog is taken to the neighbour's home where castration is done before it is released. Immediately, the dog would run back to the owner considering the neighbours as enemies. I finally concluded that the western countries have treated the murder of our people similar to the dog situation. I argued that such moves must be appreciated. That what is going on is legal castration not the use of knife. There was applauding from the worshippers before I gave the floor to the church secretary. Outside the church, I met some old friends and we had a chat. We took a photo viewing Nile River together before I proceeded to my brother's home.

After two days in Bor, I took another bus to Twiland. I arrived at Wangulei at three pm. I walked for ten minutes before I arrived at our ancestral home. Laid bare with short grass was the land we used as a farm. There were group of old women including my stepmother waiting for me. I was warmly welcomed with prayers. Then I was taken round the old cow byre where I was asked to locate where my Dad was buried. I did. I sat down and while looking around, I broke down in tears. I met my mom and stepmother Nyanwut.

Figure 19 our cow byre 2012

There were many people I could not assist with some in need of medicines and even food to eat. There were little children who came around. I tried to share what I had and it was not enough. The women began to comfort me and asked why I was emotionally broken. I said nothing. I was broken by the level of development the country was still in starting from Jemeza where I found many children boys and girls selling small fish to passing passengers by the road side. They were among the millions of children who missed schools every year across the globe.

The next day, I moved to Pongborong to chat with the teachers. The school was built by villagers out of grass though Ayual community Development Associations in the United States, Australia and Canada and the South Sudan branch had initiated the construction of permanent concrete walls. I met with teachers and interacted with over hundred children in overcrowded dirty rooms.

Figure 20 Ayual Community 8 September 2018 fundraising Committee for Construction of Maternity Ward from left, Simon Yak, Achok Mayen Jurkuch, Bul Deng. Atem Y. Mayen, Mayen D. Juach, Mayen Ajok, Chol Yaak Akoi, Arok Bol Arok, Deng Yaak, Akoy Machok, Nyankiir Bok, inter alia
(Maj Gen Garang Akok compound, Juba)

ABOVE: Figure 21 Deng Dr Biar announcing his financial Contribution as special guest, 8 September 2018

BELOW: Figure 22 Ayual Paramount Chief Bol Manyok Duot, Sub-chief Chol Dau, Former Twi East County Hon Dau Akoi, Ayual interim Juba Chairman, Abuoi Atem receiving chief guest Eng Paul Adong Bith

CHAPTER 17

Ponborong Primary School

It was the end of the term and certificates were being given to students. There was one little girl who had been coming first in class. She was intelligent. I was given a chance to speak and I encouraged them. I then gave some money to the little girl. I left the school with promise that if donors under one generation could do something, then that would go to them.

At Wangulei were a group of Nyuak soccer players with Manyok Adoor Akechnial as their leader. I promised that we should send them shoes and sportswear something we later committed ourselves to do under the Rights to Sport Fundraising organized by Prof Jim Crockery though the Kenyan customs authority later confiscated them notwithstanding a written letter to them and copied to our supporter Bishop Isaiah Majok Dau expressing that they were charitable and not chargeable under International and Kenyan domestic laws.

I spent two days between Panyagor and our small house built by my step mother at Wangulei. I spent about a week there and took off for Juba. I went to Panyagor on a motorbike where I took Landcruiser and we arrived in the evening in Bor. I spent a day there then the next day I met a group of passengers going to Juba. I recognised Bul Juach whom I had met a lot in the 1990s. We took the front seats. I sat next to the door and he was near the driver. Upon arriving in Malualchat there was a convoy of military vehicles recklessly moving fast.

Then I decided to remove my camera and upon seeing the camera they suddenly stopped as if they had found the one who had been encroaching on the South Sudanese territory of Abyei and its borders with the neighbouring countries of Kenya and Uganda.

Every passenger got scared except Bul. I was pulled over and slapped by about six to seven men. I could not even recognise anyone other than the scarification on foreheads. They were soldiers of Division 8. I was taken inside a vehicle fitted with a 12.7mm machine gun. The gunner was mercilessly kicking and putting pressure on me. I was telling them that this camera was something else. They took the camera and unable to operate it they gave it back to me to unlock it. I did it. They went through the pictures against the backdrop of noisy passengers booing them. After some minutes, they released me. By three pm, we arrived at the Bor bus station before I crossed the Juba bridge to main town.

On 22 May 2009, I left the Juba International Airport for Nairobi. I spent a night at Bishop Isaiah Majok Dau's house in Nairobi. Then I proceeded to Karatina to see my uncle's children. On 25 May, I left Nairobi arriving at Brisbane International Airport on 27 May 2009.

I was three weeks late for my classes at Bond. I had three semesters left of my degree. As usual, I continued to meet prominent personalities and students coming from diverse background. The graduation was still a year away. I prayed that God would see me through.

In one of the equity classes, I met a committed student from Zimbabwe. She was good at equity and she understood the principles put forward by Prof Denis Ong. We found a reading group. This equity law had calculations in constructive trust and equitable interests. In April, 2010, I was informed of my graduation day on 20 June 2010 for the Degree of Bachelor of Laws with Integrated Legal Skills Certificate.

The graduation was what I had been waiting for and I explored the Brisbane malls for expensive suits. I made sure everyone was informed to include my brothers, uncles, friends such as Ayiik Garang, John Awai Piok, Garang Ayiik and Rev. John Adoor Deng as I had spent most of my time in his house free of charge. By ten am, the graduation was almost completed. Barrie Hansen provided what I needed that time as a friend.

I met Barrie Hansen in a company law class. He was doing his Juris doctor. He was over fifty by the time I met him. He has vast experiences in the oil and gas industry and in road construction. I also took the Energy Law class by Prof Tina Hunter with him. One day we went to lunch where we discussed several issues from wars in the Sudan and to oil and blood, the impact of China in the horn of Africa and western involvement.

He drove me through the Gold Coast and around the beach. Andrew David and his wife Kelsey Andrew who were also students at Bond were taking group photos. This reminded me of the documentary we ran with Michelle in 2009 on the Lost Boys of Sudan under Suffering, Hope and Faith. We sat in an orderly manner and our names were called out by the faculty. Two of my best friends Kate Michelle and her boyfriend took the stage for academic excellence. Like what they did at The Hague, they made their parents proud a second time. We took group photos with Vice Chancellor Professor Robert Stable, Prof. Jim Corkery, who was associate Dean of Bond Law School from 1988 to 1993 and Law Faculty Dean Geraldine Mackenzie.

The family was joined by other renowned Professors. Then I and uncle Malek's family, my brother's sons Alaak and Jok and friends drove to Robina Town Centre for lunch at twelve pm, and finally to Surfers Paradise. I was still working under Barrie Hansen as the Southern Sudan Relations Director for his company, Eden Petroleum. Barrie had good intentions to help Southern Sudan to become an independent state.

With his highly demonstrated commitment to our peoples' cause, I introduced him to Sudanese community in Brisbane and few politicians who came from Juba, South Sudan. Some of my friends like Kur Ayuel and James Kai began to organise workshops on referendum preparations. I came a long with Barrie who amused everyone on his knowledge of Sudan political dynamic problems and how Southern Sudanese assistance should be of international concern.

The next day, we were invited for dinner at Dr. Luka Biong's house. There was enough food and wine prepared that the winds of our freedom were visible. Luka took us through his experience in East Timor, a country which seceded from the powerful state of Indonesia on 20 May 2002. East Timor is an island at the southern end of Southeast Asia was like Southern region of Sudan. It was under the control of Indonesia. He said it suffered extreme violence between separatists and Indonesia Military.

He requested that Barrie make more connections with the Southern Sudan Mission to Australia and New Zealand Marino Deng Ngor who was based in Canberra. We stayed there until one am because Barrie was going back to the coast. We left a few individuals over there to finish the remaining wine. We took his contact details and promised to come back again.

To assist in South Sudan's internationally monitored referendum scheduled for July 9, 2011, I booked a plane ticket by Southern Sudan Mission to Australia and flew to Canberra.

I met Marino and his three staff and we charted the way forward since the Referendum allows all the Southern Sudanese in Australia to vote. There was another political level to the exercise.

Youth in the country were also mobilising everyone to exercise their international constitutional rights to determine the future of their country. By mid July, I was asked by Eden Petroleum to promote their oil interests and investment in Southern Sudan in exchange for building roads, health and agriculture sector. I booked the ticket and I left Australia to Juba via Dubai. As usual I was picked up at the Airport and taken to Juba city for accommodation. I rested for two days and began talking to government officials.

I first started with Commander Chagai Atem Biar, a beloved leader who indirectly helped me out in the mid 1990s with the boat I used to cross the Nile to Bahr el Ghazal. I told him that our allies the western investors wanted to invest in Jonglei State, Block 5B. His response was that Southern Sudan sits in untapped oil reserves and resources in the region and the investors are highly encouraged to develop the country. He gave me his contact details and urged that I should come to his office anytime. I was provided with cold water. He was excited by the very seeds of their struggles; the lost boys many of them child soldiers had acquired western education and technology. Luka was in Khartoum on a government mission. I rang him while I was in Juba and asked him to approach Dr. Lual Achuek Deng on my behalf.

His response was positive for he informed Dr. Lual about it. He was appointed the Minister of Oil in Sudan's Unity Government. He is from Jonglei State and Twi East County. He is from Nyuak Payam as me. While on the phone, Dr. Luka suggested that I should go to Khartoum, but I wanted to stay in Juba. Then he said that I should meet Dr. Lual when he came back to Juba for government briefings. A week later, I met Hon. Garang Diiing, who was the Minister of Petroleum in Juba. There were two governments: the one run by the Southern Sudanese and the Unity Government which comprised everyone in Khartoum.

I met the Minister before he referred my proposal to his legal advisor Counsel Isaac Yak who works there on the secondment of Ministry of Justice and Constitutional Development Affairs. I was advised that oil contracts were non-renegotiable and any changes would likely take place after independence. As required by my job, I briefed the management in Australia on the meetings. The next day I asked to be connected with Governor of Jonglei State, with whom I had a brief telephone conversation earlier. He advised that I should direct all my inquiries to the two ministers for oil.

My appointment to meet Dr. Lual was made possible by Hon. Jurkuch Barach Jurkuch who had his cousin Lawyer Jurkuch from Canada. I was asked to attend the dinner organised at Juba Gland Hotel. The dinner was attended by former Anya Nya One leader and presidential advisor Joseph Lagu and his other senior friends.

I learned a lot from the man who fought Arabs in midst 1960s. Gen. Lagu was powerful figure in the Sudanese Military and was a commander of Anyanya. It was a rare opportunity to ensure that the rules of evidence were respected. I asked for permission to take a picture with him.

As a soldier, he was still very strong. At about seven thirty pm, we were joined by the rest of the senior officials. I was asked to order any type of food of my choice which I did without hesitation.

After the dinner I briefed the Minister of Oil on our projects and the willingness of the western countries to assist with the referendum. He advised that I should see Dr. Kon Bior the renowned commissioner of Oath and a former legal advisor to the late Dr. John Garang de Mabior.

The next day, I went to see him. His office was next to the Quality Hotel. He advised I should register the company while taking into consideration that the oil contracts with Asian Oil Companies might not be terminated other than applying the principles of territorial International public law on the transfer of rights to the seceded state.

Again, I briefed the management, and they said we should wait until the Southern Sudan political situation became clearer. Then on one Sunday I went to the Juba Grand Hotel to have tea. Then I met the Deputy Speaker of Southern Sudan National Legistrative Assembly, Retired Lt Gen Daniel Awet sitting with his son Manyiel who came from Sydney. It was an opportunity for me and so I made a quick introduction and briefed him about the western investors interested in doing business in South Sudan.

He ordered the waitress to provide me with anything of my choice and immediately I ordered a coffee with milk. He gave me his business card and asked that I call him anytime and said I should come to his office at parliament and after three days I took a motorbike to the old parliamentary buildings constructed decades ago. I was cleared by the security at the gate and allowed to proceed. The motorbike rider was not allowed and I told him he should leave. Then I passed the office for speaker of SSLA Hon. James Wani and got lost and ended up at Hon. James Wani's office manager's office.

The manager was nice to me when I introduced myself and that I was coming to see the Deputy Speaker. His reply was that the deputy had gone for an important government meeting and that I should find another day.

I became so busy that I did not go back but I still remembered Lt Gen Awet in August 2010 when I met him for the second time in Hon. Deng Alor Kuol's office. I was in a very long queue and wanted to meet him as I was introduced by the former envoy to Australia Mariano Deng Ngor who was later promoted as the South Sudan Ambassador to Kenya. There were many dignitaries waiting to see Hon. Deng who had become the Southern Sudan Foreign Affairs Minister replacing Dr. Lam Akol.

Having spotted me in the midst of the people, Hon. Awet sent his bodyguard to call him. Without delay, I immediately joined him while talking with Hon. Deng. He introduced me to Hon. Deng Alor and said that the Minister should assist me for the project's success. As a result, Alor gave me his number. I also took his business card and after a week, I made a phone call with the intention to come and see him in his office. Southern Sudan was a few months away from the referendum and government officials in the Foreign Affairs Ministry were busy meeting foreign diplomats. I continued to make appointments with many government officials with the majority of them making promises which disappeared in thin air in less than a week.

Juba was small and dirty town by then. At about three, I went to the Central Pub Hotel which was located near Juba's judiciary complex. I met many government officials and military officers passing their leisure time with various games including playing cards and dominoes. Then I was again introduced by Hon. Jurkuch Barach to Professor Madol and Professor Ajuei Magot.

Sitting next to a grass thatched house was retired General Majur Nhial Makol. I had heard about him a few years ago in Kenya when I was in form one with his son Nhial Majur. I wanted to learn from his guerrilla experiences. Majur was a senior commander in the liberation movement. He joined the rebel movement with the rank of Lt Colonel making him more senior than some of his comrades.

Their seniorities, he admitted, had nearly caused some of them trouble. He said that the liberation of Southern Sudan shed more blood and that he has been retired, he was certain to see a beautiful country in one year time when the referendum came. He further alluded to me that the time would come where young people would take the country to sustainable peace and prosperity. He continued that their time as soldiers from the Sudanese Army to the rebel movement were the toughest though they had to accept any consequences to free themselves, their children and grandchildren from the bondage of oppression to build a free and fair liberated New Sudan.

I spent four months in Juba meeting with people of diverse backgrounds from academia, legal profession, doctors, politicians, prisons, to wounded heroes, and military officers, relatives and friends.

Then I decided to go back to Australia. Instead of coming back for short time, I decided to stay for nearly three years in Australia pursuing a graduate diploma of law and a master's degree.

I arrived in Brisbane 27 May 2009 and I spent one year working with refugees as a Project Coordinator before Mariano Deng Ngor appointed me on a temporary basis as the Mission's Legal Consultant. I was provided with a plane ticket to Canberra at the Government of Southern Sudan's cost. I flew with Jet Star and arrived in Canberra in the morning before I was picked up by Adoor Wal Biar and taken to his house. I stayed there temporarily.

The legal consultant position was voluntary. I was not paid anything either by the Southern Sudan Foreign Affairs Ministry in Juba or the mission staff. Hon. Deng is a humble and kind leader. He had lived in Panyagor during the 1970s where he witnessed the impacts of the Jonglei Canal project on the inhabitants of the land.

He introduced me to everyone including the security guards at the gate. He walked me around the mission building. After a month in the job, he called me to his office. He removed an old magazine with a column written by late Dr. John Garang de Mabior's tutor at the time when Garang was in Iowa. He wanted me to read through it. I read it many times. It was a professor who was stunned by the intelligence John had with his outstanding academic records and leadership. This is what was described: "John is genius, an intelligent person who will do a tremendous job for his people in Africa and beyond." We finished our tea talking about Garang's legacies and how his leadership would be missed after his death.

I spent few months before I asked Mariano to allow me to go back to Brisbane. I told him that since the Ministry of Foreign Affairs in Juba have decided not to respond to my recruitment request, I would not provide pro bono services to the government anymore. I said to him, the Sudan People's Liberation Movement and Army should not take volunteers for a ride when officials are buying the most expensive cars in Juba.

Then I said: "You know Ambassador, twenty one years of armed struggle were based on voluntary service and some of us had contributed enormously. So I needed to be paid."

He was like a father to me and instead he smiled. Then I told him an old story that once upon a time in certain village in Sudan, there was a man and his wife. They had children and they were around the fire during the cold month of July. There was no food, because there had been famine in the area the previous year. The family had nothing. The man was nearly fainting and his front teeth nearly caught fire.

The wife said: "Why do you want to burn your mouth?"

The husband rudely replied: "What would I do with the mouth?"

"There would be plenty of fish from the Toch wetland and you would be full." his wife encouraged him.

"Who would bring it?"

"Your brother in law." his wife continued.

"Would he bring readily cooked fish?" The husband hopelessly continued.

"Yes it is possible."

When the SPLM/A is a ruling party, we the supporters are like that man. Our mouths would be blown up by hunger before all the honeys promised during the liberation reached the followers in the villages and beyond that even here at the mission. The story was interesting. He was amused.

His last reply was that he was working out my employment issue with the Foreign Affairs Office and that my travel to Brisbane was granted. By eight thirty am, I had arrived with a taxi. He ordered his secretary to pay the taxi fare. I went directly to him and after some discussions I took the cheque and I left for the National Bank in town. I cashed in the money.

Five days later, I asked my cousin Adoor Wal Biar to drop me at the airport.

Back in Brisbane and Toowoomba, there were a growing number of African migrants and I thought it would be a great idea to assist the refugee communities on a pro bono basis in Queensland. I found John

Adoor Deng and Chol Aleu Angok waiting for me. I got deeply involved in community social activities in Brisbane and Toowoomba including church activities. We were the group collecting contributions for the house of the retired Bishop Nathaniel Garang Anyieth before he paid us a visit with his wife Mama Atong.

Some of the activities would later include African cultural activities. Now, the days of the referendum were approaching and the Brisbane cultural centre was a voting booth. I registered and the Sudanese Leader James Kai who later nominated that I should volunteer as the community social affairs secretary. I agreed and as a result we promoted many activities before we recruited more people to work as observers during the referendum Educating voters was occuring across Southern Sudan and designated countries where Southern Sudanese were allowed to vote. We were doing ours in churches because the population was concentrated in Brisbane and Toowoomba and required little resources to conduct major training. Also, a majority of the members were literate compared to those in Southern Sudan.

Involved in the process were internationally respected personalities such as the former U.S President Jimmy Carter, former United Nations Secretary General Kofi Annan and former Tanzania Prime Minister Joseph Warioba. They were observers. Ours in Brisbane was being run by the Southern Sudanese community and a few officials from Australian Electoral Commission. It took six days from 9 to 15 January 2011 before the Commission Head Justice Chan Reech Madut announced a victorious vote of 99.57% for separation.

Ours in Brisbane was almost 99.8% before it was sent to the polling station. Then the first violence which had rocked the country since the colonial periods was ended at ballot boxes although the Independence Army in the South was ensuring that violating the agreement was severely responded to.

The United States administration like they did during our darks days in the 1990s was standing with the people of South Sudan. Sudan known as a sponsor of international terrorism had no option for the invasion of Iraq by the former George Bush administration had scared them. For in most of international conferences or United State homeland politics, President George Bush Jr would appear on television referring to some countries such as Sudan, North Korea and other hostile countries as the axis of evil.

CHAPTER 18

Darfur Conflict and Involvement with Save the Darfur Coalition

SUDAN IN THE WAKE OF 2003 had a bloody conflict in Darfur which was reported as a genocide. To find an international forum to advance the flight of Fur people and other African tribes in Darfur, I sent an email to the Save the Darfur Coalition to be included as one of their activists. A week later I received an email and we began to send several petitions and letters to the U.S Administration. Sometimes I would make phone calls to the White House Darfur desk using the hotline which was directed to the communication team in the White House who would later relay the request to the United States President George W. Bush. In our petitions, we would urge the administration to take tougher actions against Sudan government for violating the comprehensive peace agreement provisions and for engaging in ethnic cleansing in the region of Darfur.

In the Sudan old policy was connected to the extermination of the black race and to convert the moderates to Islam. This ideology was fuelling the war and some of my students with Darfuris in Queensland had to organise a march to the Queensland Parliament for international actions. I had megaphone with me. I shouted louder with the intention that the Queenslanders would join our team.

To my surprise not many would buy the idea that there were people in

Western Sudan becoming victims of daily persecution and displaced from their farmlands. Everyone including extreme white Australians passing by us would just look at me and mind their own business. Before I called for water to soothe my throat, I was like these people have their own genocide of Aboriginal Australians in their back yard and just wanted to pretend that genocide did not exist for fear of being embarrassed about what their grandfathers did to the true owners of the land.

We did this for an hour under the banner of "No to Genocide on our Watch" before we dispersed. The only place where we could see Aboriginal and Torres Strait Islander peoples being properly remembered were some functions organised by internationally renowned scholars. Some people with good hearts would recognise the owners of the lands by mentioning the tribe that had already been forced to extinction in an acknowledgement of country.

The history books were suppressed to not reveal any evils of the past until 12 February 2008, when Prime Minister Kevin Rudd made an official apology on behalf of his government for the Stolen Generation, or the removal from their families of children of Aboriginal and Torres Strait Islander.

It was purely a political apology without substantial compensation for the victims as I was watching it on the television at Sailfish Cove.

With the Independence of South Sudan on 9 July, 2011, the war shifted to the Nuba Mountains, Blue Nile and Darfur. The Nubians were unquestionable loyal fighters in the SPLA/M; they did not backstab the movement as the SPLA/M leader referred to some southern commanders in the peak of the 1991 split. There was a desire to see them as free citizens with equal rights in the Sudan. Abandoned and betrayed, some of these comrades I met at Moorooka a southern suburb in Brisbane, had no form of regret. They remained proud Nubians, people whose kingdom dominated the ancient Kingdom of Merowe as far as Egypt.

In the Queensland Sudanese Community Association formed by Sudanese themselves were two camps, the extreme divisive figures who want independence celebration organized without the Nubians and people from the Blue Nile left out by the outcome of referendum because of the borderline of 1956 which had the two regions of the South Kordofan and Blue Nile states in the north. The popular consultations on their choices of governance which were provided to them by the comprehensive peace agreement provisions was violated and not implemented.

There our team under Sudanese Leader James Kai based on my advice not to let the political arrangements back home divide our veterans because of the geography of the two Sudans preferred a collective and Sudanese shared independence celebration. This did not go well with the group of Danie Lee, Machuei and his associates. They had to convince Hon. Marino Deng Ngor and had their first celebration.

Some of us boycotted it because it was not meeting the aspirations of every person the movement had drawn to the concept of new Sudan. Our argument before Minister in the Office of the President Dr. Luka Biong who was willing to reconcile the two groups was that South Sudan, which was now liberated, had for its freedom shed the blood of unsung heroes from the south to the north.

Figure 23 John Garang and Salva Kiir with SPLA officers during 1986 Buma Offensive being directed to the enemy line by Aguer Manyok Aguer before he was killed at Buma Garrison. Courtesy of Eng Deng Diar Manyok

The file of the liberation had people across Sudan including Arabs who hate oppression such as Yasir Arman and Dr. Masour Khalid. The Nubians required being treated as first class citizens for many of them I knew in the past had died in the many battlefields of Southern Sudan and the Nuban Mountains regions.

I had interacted with some of them during the war in the midst 1990s and their contributions to South Sudan by scale is far beyond the contribution of those who considered themselves as genuine south Sudanese. In the mid '80s and beginning of 1990s, there were ambush related deaths caused by South Sudanese themselves.

In that Brisbane meeting, there was no agreement to bring the two groups together. The mediation was polluted by a bitter exchange of words that one of the Nuer elders and educated Samson said that one cannot attend negotiation and go back home with 100% otherwise such negotiators deserve not to live in an island not lived in by quarrelsome human beings.

Dr. Luka's skills of mediation did not work and he left. A week later we mobilised resources through the Sudanese Community Association in Queensland to have another celebration. We then sent invitation cards to all those from the south and the north. This celebration was in great contrast with the one organized by the newly formed South Sudan Council of Communities under the likes of Daniel Lee and his executive.

In a meeting organised at Moorooka, I was nominated as secretary and at the same time the Master of Ceremonies. We invited everyone including the original Arabs since it was necessary for them to witness the new nation South Sudan to talk of their experiences.

One of our notable speakers was Dr. Luka Biong. He sat in the front row with the First Secretary from the South Sudan mission to Canberra. The celebration as being common among the Sudanese Christians started with prayers. Then it was followed by an introduction of the programs before I briefed the audience on the activities and speakers of the day. I made an introduction and informed the audience that with independence our passports would not have restriction to any country even Israel. I continued that one could use it everywhere one goes in the world. Everyone applauded including Dr. Luka. Dr Luka who was among students Sudan knew very well how Arab-Israel bad relations became a national issue to both Arab states which considered Israel as an enemy of Islam following the Six Day War.

By eight pm, the celebration ended with dances from Acholi and other tribes. The event was filmed. The prayers concluded the celebration and we drove off to Toowoomba.

I stayed in Queensland for some time and now I had to enrol in a master's degree in Law, Media and Politics at Macquarie University having declined the Master of Law offer from the University of Canberra. Then, I changed my mind and I enrolled with Master of Policing, Intelligence and Counter-Terrorism at Macquarie University, Sydney.

When this master degree was concluded, I enrolled full time and completed the practice training for the Graduate Diploma of Legal Practice at Bond in 2012. I did it for four months full time until the end of April under intensive assignments and examinations. Then I was

Figure 24 Macquarie University ground during the graduation 18 April 2013

connected to Boscher Lawyers by my former lecturer, Joseph Crowley, where I practised for a short time. While there, I had to appear in Brisbane Courts under the supervision of senior Lawyers. The firm specialised in crime defence cases of murder, drug trafficking, robbery/burglary and traffic offences.

The law firm began to refer cases from people of African origin to my desk though I was still under training. I had a good supervisor John with me who was once a former student at Bond. He assigned me two cases, one was for a stolen car and the other was concerning driving without a licence and under the influence of alcohol. In both cases, I had to seek court leave to appear as I was not allowed to practice until I got my license from the Queensland Supreme Court.

I was lucky that the two offenders received lessened punishments. The traffic offender had eight counts on her charge sheet with a clean criminal record. I drafted her guilty plea and took a train to 43 Ellenborough St, Ipswich where the court was. I entered the gate where security did the screening and I was allowed to go inside the Ipswich Magistrates Court. The offender was a single mother of six children. She was driving to the market to buy food for her children. I read her plea and the female judge accepted the accused's mitigating factors.

She was fined $800 payable every month for three years. This was a reduced fine from $8000. In addition, she was asked to do more practice before driving again. Everyone was excited by the court outcome and I immediately took a train back to the office. I briefed John and he was mesmerised. The other experiences were serious cases of drug dealers who had killed one of their friends and dumped the body between New South Wales and Queensland. The police later discovered the body after a series of serious investigations. The fugitives were interrogated and recorded. They had to locate where the body was under no threat of torture or otherwise. During the trial, the main accused entered into a no guilty plea therefore attracting the empanelling of jury in accordance with law on the right to be trial by jury and common law presumption of innocence until proven beyond reasonable doubt. The case dragged on for several months with the calling of several witnesses including a DNA analyst from University of Melbourne.

It was already May 2012, and I felt like I had to look for better job at Darwin, which had the Aboriginal people of the Larrakia language group. These people were its first inhabitants.

The place was named after naturalist Charles Darwin by Ship Captain John Clements Wickham. Historians in our high school had it that Darwin natural selection includes the survival of the fittest. Charles and Wickham were thought to have sailed together during the earlier second expedition of 27 December 1832 to 2 October 1836.

Waiting to pick me at Darwin Domestic Airport was John Awai Piok Keer. Awai is an honest and kind gifted young man who had accumulated life experiences from church and virtuous upbringings. Whenever we met, we would spend the all night deliberating on important matters to our lives.

There were times we had to spend days in Brisbane at the house of Abuoi Piok Keer. He drove me to Coconut Grove for lunch. We sat watching fish bouncing up and down from the Timor Sea. In Darwin, I met the likes of Lual Mayen Dhieu, his brother, Isaac Thok, Wach Abit, Garang Malual Jok and his family. We had a good time together and we could explore Darwin Harbour which was bombed by Japanese forces in 1942. I was taken to where the Japanese destroyed parts of Darwin. Deep inside Darwin's forests were hazardous remnants of World War II and extreme care had to be taken by those who wanted to walk in the bush.

Dhieu Magot was a trained Australian soldier in the reserved list who led us around the bushes during the next weekend. He would say people still have a stigma and fear of being exposed to ordinances still deeply embedded underground despite advanced technology. Then I got a job in the Darwin Detention Centre with the connections of Awai who had worked there for a year. The centre was a confinement site of asylum seekers interdicted at sea and in Australian waters. These vulnerable people were detained without legal due process before the migration department look into cases.

There I could interact with many the traumatised victims from those who experienced civil war in Sri Lanka to those escaping the effects of the Indian Ocean Tsunami of 2004 and their stories nearly made me cry.

Those who could not withstand extreme isolation with no prospect of a bright future committed suicide. The level of unlawful confinement that these people went through was beyond the standard of Australia which is often praised by western media as one of the strongest economies in Asia.

I worked there for only few months until December 2012. Then I disagreed with my inconsiderate boss and I resigned without giving any reason in writing. I spent one month without a job and got another job at the Defence Establishment Berrimah and RAAF Base Darwin.

Here I met new friends including former navy officers. I was tasked with ensuring that the swimming pool used by soldiers was well guarded. It was a security related job. No skills were required but the job was enough to help me pay my bills. Two months into the job, I was called for an interview for the position of Assistant Lawyer in one of the Northern Territory law firms with an office in Darwin city. I dressed well and attended the interview. I presented my academic credentials. I met the requirements of the job according to the testimony from the young lady on the panel.

I was informed by phone that there was a possibility I would take the job in two days' time. I waited for more than two weeks and I knew something was wrong. Attempts to call the law firm were not responded to. Since lies and job discrimination remain dominant among money making corporate bodies, I firmly stuck to my job at the RAAF base in Darwin.

I spent the first quarter of 2013 in Darwin. I got out of it to attend my Master's Degree graduation at Macquarie on 18 April 2013. I spent the night at my nephew Mapiou Mun's house in Sydney before he drove me to school to attend the graduation. I was joined by my friend Chol Ajak Ariik, who had conferred upon him a double Masters in International Security. Ayen Marol Juach and Makuei Ajak Arrik among other relatives cheered for us.

By the time I and Chol graduated, we had mastered security related threats and countering measures not only in Asia but across the globe from terrorism, insurgencies, non-state actors and security intelligence analysis and organized syndicates' mode of operations.

Mastered with skills on human safety and experts on terrorism and counter-terrorism, we made another life changing decision to go back to South Sudan, the country where millions of lives were lost and blood shed to assist in nation building.

After few months, Chol and I left Brisbane for Juba. He met some of our colleagues already working with the government to advance the quality of life of the people. Their founding fathers had prophesied that their own children would provide them with adequate services when they returned back from exile in what would be a promised land. A land of flowing honey and milk.

Juba by the time I was there in 2010 was home to both vultures and honest investors and very ready public money launderers. It was a region with emerging new institutions established from zero and rejuvenated old buildings. The laws on financial accountability were still being developed. Then the smart investors saw loopholes and began to rush to Juba before 2005. Some of them were already in town like Yei and Nimule while praying for SPLA Commando units of officers such as Maj Gen Atem Aguang Atem to bravely and forcibly enter Juba from the main Bridge.

The commando entered without resistance from Sudan Army Forces and everyone was busy mobilising for Southern Sudan to make money.

With money flowing from oil and non- oil revenues, few of the very men and women who fought the injustice of corruptions and inhumane treatments were practising both with impunity. There were also investors of all types including East African convicted jailbirds in Juba. Night robberies with violence began to increase and one could not go outside past midnight without major concern for one's own safety from family members.

On 15 July 2013 in the evening, I called Awai, Garang Malual and Dhieu Magot to get advice from them. I told them that I wanted to move back to South Sudan now a country with three years of independence. Without hesitation, all of them said that home, though short of all goodies of Darwin, is life without stress and cruel riches driven by capitalist corporate bodies. They said at home, there would be no phone call waking one up before breakfast for an urgent payment of last month's electricity bills. They all supported the idea.

In the major cities of the western world, there is no paradise as is often perceived in the developing world. On the streets of major cities hidden in dark corridors are both men and women who are reliant on dangerous drugs. Having been overpowered by the abuse of a substance, some will later become homeless and will sleep underground in deep tunnels. Their frustrations would later include harassing passersby with abusive words such as black Negro and question why one had to be in their country.

These street drug addicts would not understand that it was the Europeans who first invaded Africa in the mid 18th century without anyone's authorisation. And that the Africans who are in Europe, America and Australia work to empowering the white men economies.

According to Prof John Wade, the problem with human beings lies in their memories. They are like chicken soup. On the way to Blacktown in Sydney, I would hear words like "What are you doing in my country, you black man" echoed by those who had chosen the street life of taking drugs. In the west as in any developed economies, what pays one's bills is hard work not the skin colour. The colour of our skins is external and inside is human flesh.

But the poor people do not discriminate by race when asking for assistance. I remembered this in Holland when I was approached by two black and white men asking for financial assistance.

They paired together and lived in harmony with one another. There was no doubt these two gentlemen knew that human blood figment goes beyond racial disposition. Like in Africa, racism though practice by few wicked hearts remains a human weakness.

CHAPTER 19

Back to Juba for the third time

Having already secured my ticket to Juba, Awai called for mercy and a safe journey via travel prayers. He did it like how my parents used to do it back in the village. In those days, if one wanted to travel to Malakal and distant lands, rituals would be performed.

Awai and his cousins wished that I would go in peace and promised that we keep in touch. For nine months after I left Australia, Awai had been in touch with me. This time, he helped me massively by contributing finances toward my marriage to my beautiful wife Awel Kuol Mayen in April, 2014. His contribution in terms of dollars together with that of Professor Laurence Boulle and Peterson Mwanga was equal to over five cows.

On 21 July, 2013 in the afternoon, I left Darwin International Airport for Juba via Milan. It was the longest flight with over 20 hours stops over in the Philippines. I could view the seas including the Java Sea from the plane flight data.

Two days before I left Darwin, I conducted a google search on the security situation of the Philippines which was prone to terrorist attacks from the Muslim separatists, kidnappings, and sometimes civil unrest.

As a visitor, I wanted to make sure if I booked a hotel outside the airport, what the guarantee of my personal safety would be Google stated that foreigners might be drugged with sleeping pills before they were robbed of their valuable items. Like in Brazil and Italy, the organized criminal

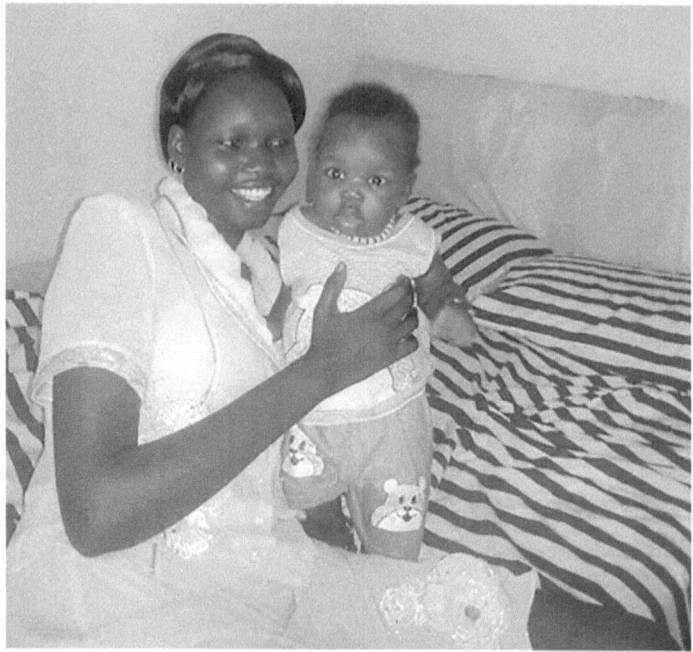

Figure 25 my beautiful and caring wife Awel Kuol Mayen aka Mama Adhieu

syndicates I studied back at Macquarie had vicious bosses, soldiers, drugs lords, which could exert considerate influence to enormous moderate affiliates. With that in mind, I decided to remain inside the airport until the Emirates airplane arrived in the morning. It was already 22 July. Then I arrived at Dubai after a long flight and then finally to Juba on 23 July 2013. I was picked up by my maternal uncle driver Ter Malual and taken to his residence at Juba town centre.

While watching the evening television news, the hotel I had gone to have dinner in was packed with neatly dressed security personnel.

Juba by then was in political turmoil. On 23 July the whole cabinet was stood down. There was wrangling within the leadership on who should be the flag bearer in the upcoming election in 2015.

Such behaviours were amounting to acts of indiscipline and the cabinet was sacked. In traditional African society, one could reign as a paramount chief until a new heir was born.

The leadership challenge in fragile post-war countries as I studied in the insurgency and counterinsurgency classes is that any democratic process in South Sudan may take time. Those who had the guts were crossing unexplored soccer field. South Sudan was where politicians had begun to taste the privileges of being leaders since we had known nothing for over hundred years other than the foreign undemocratic dispensation.

The next day, I moved to my uncle's place near the Army headquarters in Newsite, where I spent five months. It was already December and everyone was looking to celebrate Christmas with family members. I had a plan to go to Bor together with family members of Alaak Jr, Ayom Diing Wieu, Jok-mapeace, Adhieu Alaak and Manyok Adoor Akechnial. The political fever was high in the country and war was imminent. The media rhetoric of the leaders was greatly exposing thousands of 2011 independence referendum voters to life endangering possibilities. The fear in Juba was not because of Omar Hassan El Beshir Mujahideen fighters, but because of the conduct of the very men and women who liberated the country and promised heavens and a land of milk.

I spent the whole of July trying to get used to Juba weather. On 8 August 2013, my cousin Akoy Machok Biar, who was working as a personal secretary to Hon. Minister Stephen Dhieu Dau Ayiik booked an appointment for me to see the Minister. I was looking for a job having made up my mind to stay in my country of origin for a considerable period of time. This meeting came after I had a consultation with Dr. Kon Bior Duot. The late Kon was an UK trained barrister and academic with more than seven degrees. I had to consult with him on general advice including the requirements for me to open my law firm.

His advice was that I should first seek employment with government to build up my experience on South Sudan's new governance and practices. He added that after I have built up my network of people, then I could open my law firm. To find a job with the government became my first priority.

By ten am, I met the Minister. I presented him with my academic credentials. Then I told him I had worked with Eden Petroleum as South Sudan government relations director, researcher, and legal consultant and project coordinator in Australia and had once been an observer at International Criminal Court. And that I had my practice with Bossher Lawyers. Without questioning the region I hailed from, he called his secretary. Then two letters of recommendations were written to foreign oil companies. They were instructed to call me for an interview if any relevant positions become available. By lunch time, the letter was ready to be taken to Eng. Paul Adong Bith the then Managing director of Nile Petroleum Company (Nilepet), a government owned national company. He took the letter, having gone through its contents, he instructed that I should come back as the matter would be referred to Baak in the Human Resource department.

The next day, I was told that I would have a contract as legal officer. This contract was to become effective in January 2014. The other letter was sent to the Ministry of Interior. Before I left Australia, I had several phone consultations with Lt Gen Mayom Deng Biar for I had said that my second option other than legal practice was to work in the judiciary or the Ministry of Interior. I came to know him in Western Equatoria in 1992 when I used to feed his tailless monkey.

He is a valourious and noble outspoken officer. He was the Director of Traffic and a respected and nicknamed as abugowie (the strongest). He would conclude that the liberators are not rich. They are only rich in legacy but not in material wealth.

He said if I came to Juba, he would take me to the Deputy Inspector of Police Lt. Gen. Kuol Nyuon, a promise he later fulfilled.

He wanted me to be part of the South Sudan National Police Force. Based on his persuasion, I was recommended to teach at Rejaf Police Training Centre an hour's drive south of Juba on Gen. Kuol's directives.

Before I left Australia, I had wanted to enrol in Police College to directly or indirectly work together with those who were strengthening the rule of law in South Sudan.

On 28 July, 2013, I rushed to police headquarters at Buluk where I met Col. Alier Apollo, father of my schoolmate late Counsel Kut. He sent me to his secretary to be provided with forms. I filled the form immediately before being informed to get ready to sit for police screening examinations in few weeks' time. I came home and shared the new development with my maternal uncle Abuoi Arok. He had reservations and advised that I should stick with my legal practise and if I want to expand my horizons, I should join the Ministry of Defence and Veteran Affairs.

He provided all the convincing justifications. He was right. He was seeing the opportunity present in working with the strategic institutions of the government. For the last three decades going back to 1989, being a soldier has been passing right by me and has never happened. For instance, my training at Kalacha did not take me to combat. I was a potential soldier in wait at many various rebel areas. In the Army headquarters, I had my recognition as a member of the Jesh el Ahmer (Red Army).

On 13 December 2013, I was given an assignment by the Director of the South Sudan National Prison Services Brig Gen Ajak who was later promoted to Maj. General. I was instructed to review some of their financial files. This was my first job in Juba on a paid basis. It was supposed to be two to three days assignment.

To ensure uninterrupted work, I was accommodated at the prison service's cost at the Star Hotel. Outside Juba were unhealthy political debates. On 14 December, the divergent views in the SPLM political bureau were causing too serious a national rift. And by nine pm on 15 December 2013, there was shoot out inside the Giadia Army Barracks. I was called by one of my uncle's bodyguards. He reported that there was fighting inside the Army Barracks and advised us to be vigilant and that some places would be targeted.

That shootout triggered the unrest in South Sudan three years after independence, which later spread to other cities across the country. Thousands of people were displaced and many wounded or killed. Juba became a war town as soldiers ran amok to frontlines.

The wounded and casualties were from a fight of brothers, the same people of South Sudan. At Newsite, our house built by a Lebanese company and meant for senior army officers had gunshot holes. Also attacked was an ammunitions storage sending panicking residents in different directions. The civilians who had worried about revenge attacks fled to the United Nations Mission in South Sudan (UNMISS) camps in Juba while other cities such as Akobo, Bentiu and Malakal, hosted a mixed people of tribes with less fighters to protect them.

A friend of mine named Angui and who used to drive us was killed in action at the Newsite ammunitions storage gate in the early hour of December.

By nine thirty pm, I was called by the artist Aguer (Tongadit). He informed me of the saddening news of the murder of Ayual Nakura's former youth leader, Aguer Yak Mabior. Now, the foreigners either investors or those working under UN got totally confused. They would run to me but no one knew what was going on until it became clearer that the fighting was between Presidential guards. By one pm the following day, my hotel became a place of refuge as exhausted soldiers rushed in there to get water before they returned to their patrols.

As a result, there was chaos in Juba and Ajak Deng Reng offered his father and elderly people such as his cousin Ajak who came from America a car to take them to Kampala, Uganda. I and Malual were also asked to accompany them. We drove to Nimule. We met and interacted with soldiers manning the border. Our documents were stamped and we were allowed to cross to Uganda. At the Ugandan side of border, it was already late and we could not get clearance by the Ugandan Immigration officers. So we slept at Elemi.

On 22 December, we drove off and made it to Adjumani. We spent a night at Reng Deng Reng's rented house. The next day we took a bus to Kampala. I and Malual Arok spent two months in Kampala. Every day, I used to get briefed on the country's situation from friends in Juba. Attempts by Australia's government embassy in Kenya to repatriate me back to Queensland were rejected with the conditions that they should take my relatives with me as well for most of them were stranded in Bor.

It did not take long before the forces of Dr. Riek Machar in Bor, Malakal, and Bentiu among other areas were repulsed or pulled back.

Having established a centre for Strategic Analysis with senior commanders and the Ministry of Defence and Veteran Affairs officials earlier on before the outbreak of the conflict, I was called back to Juba. The senior officers of Maj. Gen Kulang Mayen and Maj. Gen. Riak Jeroboam Machuor had consultations with Lt. Gen Kuol Manyang Minister of Defence and Veteran Affairs and Chief of General Staff Gen. James Hoth Mai. I was offered a position as a Defence Affairs analyst on 8 February 2014 based on the structure of the centre we proposed on 25 October 2013 with comrade James Chol Ajak.

My job was designed to report directly to Maj. Mawan and Chief of Electronic Warfare. It was sad and heart-breaking that our people were murdering themselves following the Sudan Peoples' Liberty Movement's calamity. The crisis had genesis in politicians' serious issues on governance and mishandling of post war dynamic challenges.

The conflict was unjustifiable. It was comrades killing each other. Officers such as Brig Gen. Abraham Jongroor perished on the road to liberate Bor. It was only a few months before that I had had a lengthy conversation with him at Newsite. He was among the participants who attended our family reunion organised by my uncle in around September, 2013.

He was in Juba when his men recaptured Boma, a strategic town which was overrun by the rebel commander David Yau. Yau was against Government. Yau was protesting with violence his failure to secure a nomination for member of Jonglei State parliament during the only South Sudan ever general election in 2010.

Abraham, when I met him, did raise issues of concern on the status quo. He said that the youth should use their education wisely because they had had education that some of his colleagues did not have because of war. He said that the civil war had denied them education for they had to spend time fighting for liberation. It was a war aimed for the advancement of people's economic and other absolute rights. I was just listening as he talked. However, I had to comfort him that South Sudan was proud of him as a true liberator. And that the South Sudan flag would not be flying today if it was not because of them. None of us knew that it would be our last time seeing one another.

With the lives of civilians at stake in Bor and the likelihood of thousands of ill-trained white army fighters advancing to Juba, there was a national duty to recall him as a soldier with wartime experiences. Brig Abraham together with other senior SPLA commanders were ordered to recapture Bor. Unfortunately, his vehicle was ambushed. He was killed on the way to Bor. One of the bodyguards who is related to me who later escaped with a bullet injury to the back narrated the ordeal.

The live video of the attack and chaotic retreat began to be shared on social media. Those who narrated how they struggled under heavy fire to take the body of their commander to safety had to control emotional tears. This bodyguard I used to meet at Giadia Barracks stressed to me that he was shot from a closer range. His death devastated people who knew him and I could not believe the BBC report that he was no more. Like many other unsung martyrs, he had left behind his family.

When I passed by his fabricated house at Newsite, I remembered the man who had shed his blood to defend the land and leadership. In Africa, the liberators are easily forgotten because the cake of liberation has a curse. It is a cake which has a tendency of attracting local and international breeds of unreputable wealth accumulated personalities. Inside the Army Hospital building at Giadia is a printed plaque attached at the gate of a small ward with his name and those of Butrus Lomich, Marach Akoon, Ajak Yen. These were senior Brigadiers whose lives were wasted in the 2013 armed confrontation.

His temporary defence provided family house is equivalent to that of a private soldier's. A private soldier whose little salary is handpicked at his unit office either twice, or slightly less than a quarter a year. As oil and brand new cars flow daily, the civil servants' salaries in South Sudan take too long. Like the cry of private soldier waiting to receive that little salary, the families of martyrs are crying in the cold. Sometimes one nearly cried and wondered if his family conditions would be like that had he not died for the country. In one of Ministries in Juba, the number of V8 cars in a small space was twenty. The current price without any corrupted dealings of one V8 is equivalent to $85,000 dollars.

While at my rented apartment in Gudele One, I called class four pupil to do some simple mathematics. His mathematics was that in one Ministry alone there was USD $1.7 million embodied in appreciating assets. Near Pyramid hotel live orphaned and poor families between the Afex and the old cemetery. As we drive to our places of work, the young men aged ten to twenty one would wrestle over who should wash the car's body. Then I felt that the country shall stop buying of expensive cars instead of using money to establish a welfare system for the veterans and their families.

When a few vehicles and awards were being issued in 2019, I thought they would start with those who were physically absent from our midst so that a good precedent is set that anyone who laid his or her life for the country would see how he or she would be treated if the time of war related death comes.

I was wrong. Testimonies on how people who are presently unjustly rich themselves forgetting the rest of who were once their comrades littered Juba's tea place talks.

In those days of struggle for independence as taught at Kalacha, the delicious parts of meat were going to senior officers. As recruits known those days as Mathejit and whose rights to ask were suppressed, the ones who had to get the best treatment were the bosses. That is what our senior officer at training considered to be discipline in the Army.

The good officers would let their subordinates eat first while the power obsessed ones would eat first before asking the subordinates what they have eaten. In 2010, I have seen bodyguards in one of the senior government officials' complexes begging for drinking water when the boss' friends were busy sipping heat chicken soups inside one of the expensive hotels. Hit harder by the scorching Juba Sun and in the grip of a power inconsiderate boss, he had to beg for water to quench his thirst.

All these practices had been there even before the assassination of one of the Ministers whose former bodyguard broke into his car and used his pistol to murder him. The golden rule has a clearer distinction on what one expects from the other. That you should treat others the way you would like to be treated.

CHAPTER 20

Working in the Ministry of Defence and Veterans Affairs

Having been fully assimilated into Defence Affairs with effect from 8 February 2014, I developed a proposal on balancing the International Laws of the Armed Conflicts, and Human rights and National Security. This paper was presented to the Undersecretary of the Ministry of Defence and Veteran Affairs Lt. Gen. Bior Ajang Duot. The paper was approved and immediately taken to Gen. Paul Malong Awan Anei the former Chief of General Staff by Undersecretary Document's controller. It was properly written and I owed a lot to Comrade Benin who was the office manager. General Malong was the former governor of Northern Bahr El Ghazal who had crossed back to the Army.

With the support of people like Col Angelos Agok, the paper was cleared and we were given a green light to conduct the workshop in the Army Barracks at Bilpam. I mobilised experts on International Humanitarian law and its intertwining concepts with National Security. We immediately presented our cost for urgent payment. It was lower than $10,000 including feeding, taxes and the issuance of certificates to the participants. The request was forwarded to someone responsible for Finances in the Ministry.

In his hands, the project was curtailed. The explanation was that there was no money. I was left with no option but to order the workshop to be cancelled. Among the participants were men in uniforms who were responding to orders of Lt. Gen Dr. Malual Ayom Dor leaving behind their duties at Ayod to attend the workshop. They had to be told to go back. The cancellation of the workshop made me felt embarrassed.

By 21 October 2014, I approached the Senior Counsel Daniel Anuol Deng Kureng, a UK barrister. He was a legal advisor to the Ministry of Defence and Veteran Affairs before he started his law firm Awatkeer Law Chambers. Daniel is an intelligence lawyer who believes in the rule of law and equity before the law. His first remark was, when I met him in the Ministry, that I should join some of his dedicated lawyers and staff to run the law firm reporting directly to him.

His immediate consultation with Chol Agoot led to my appointment as a senior counsel after an interview. I was assigned to handle both civil and criminal matters before all courts in Juba and beyond. To ensure the law improves considerably, I dedicated all my time to legal research and reasoning. The law firm had grown to include young lawyers like Geu Kuol Nyuon, Atem Diing Akol, Bol Majok, and Neil Lado. We were later joined by counsel Aguer Chol Aguer and other non-lawyers of Garang Thon and Geu Manyok Akuak.

Instead of being saluted by soldiers I was dealing with five cases from both local and international clients. One week later, I had my first appearance in Court before the notable Hon. Judge Francis Amum. I was still in adherence with the use of Australian accent and common law approaches to the matter before the court. Hon. Judge Francis was honest to tell my opponent and Arab pattern lawyer that I came from other jurisdictions and that he should understand.

He said advocacy is not about handling submissions and applications to the judge only. It requires educating the judge on the theory of the case and going through the precise summary of elements one had to deal with. My opponent was a local lawyer who later became disgruntled after his client's case was dismissed. He took it personally. He misunderstood a contract which was drafted in accordance with the laws of the United Kingdom (Employment Rights Act 1996). He was trying to mislead the court by arguing using the laws of South Sudan instead of looking critically at the expressed and implied terms and conditions of contract. For me to succeed in court as Australian former judge and criminal law lecturer David Field to put it, the unprepared lawyers are empty vessels which make more noise in court. Having this in mind, I had adequate preparation.

My opponent's client's case was thrown out by the judge although he highlighted some issues I raised before the court like the jurisdiction of South Sudan courts to try a case in which terms and conditions were expressly foreign. It took Peter nearly six years before he could greet me.

My encounter with him is that he was one of few advocates who deserve professional legal training if the South Sudan judiciary wants to escape the administration of untainted justice. I said he should attend legal training outside the jurisdiction of South Sudan for a few months to make him access to deal with cases of United Nations recruitment malpractice. He does not know that legal practice is like battle. For a soldier losing a battle one loss does not make one forever the loser.

I was charging clients based on the merits of their cases and sometimes provided pro bono services to the underprivileged. With independence, there were always dynamic challenges coming with nationhood. Those challenges had to include the lack of opportunities for the youth. For instance by the time I was licensed as advocate, there were many unemployed youth who refused to engage in burglaries and had resorted to roadside money exchange businesses against the regulations of the government. While going to my office, I was approached by an old friend who said he knew that someone by the name Bul and his friends who were arrested at the Malakia police station were at a dirty prison referred to Konyokonyo for its debauched smell and congestion.

The allegation was that they were caught in dealing and their money had been confiscated by the police officers on the false pretext that the practice is illegally engaging in dollar exchange and that such people were the ones destroying South Sudan's troubling economy hence the threat to national security.

There were over 28 in number. Bul who was his group leader told me outside the prison room that their money had been taken by the Juba City Council and that they had been accused of dealing in criminal activities of illegal money exchange and therefore he was picked up by the police near the Ivory Bank. He continued that anyone with a handbag had to go through bag searches.

He sought my legal advice and asked that I represent him in Juba County Court behind Malakia police station opposite the road to Konyokonyo. I accepted representing him free of charge and on 9 February 2015 at seven am I was in the office working on his application. I drafted a three page application to the court and supported it with legal provisions. In most of the arrests, there were no charge sheets as the charges were murky. It was a random pick up motivated by various forms of pecuniary interests between the police and court officials. By the time I arrived at nine am, I found all twenty-eight defendants arraigned in the court bench. I rushed to the front row and sat next to the prosecutor. I told the prosecutor that I was the lawyer for the accused. Then the session began with the prosecutor and cross examinations.

The prosecutor failed beyond any imagination to prove whether he found the accused exchanging dollars, and what the proof was. He could not produce even one dollar as evidence against the accused.

I had with me overwhelming documents and any other circumstantial evidence including the bank receipts to prove that the accused were going to the bank and had not engaged in any criminal activities. The prosecutor also failed to produce eyewitnesses although he relied on statements from another 2nd Lt. Police officer. I immediately requested the judge to dismiss the case and order for the money confiscated to be given back to the rightful owners.

The judge's eyes were instead glued to the bags where money was kept. He brushed aside all the underlying legal processes and made an order that the money had been taken by the government for breach of local laws and that the accused were fined 250 pounds each. The amount of money alleged to have been confiscated was in the millions of South Sudanese pounds. Immediately, there was chaos in the courtroom. My response was to ask Bul to live with allegations and to consult with the most experienced commissioner of Oath Dr. William Kon Bior. The late Dr. Kon was fluent in both English and Arabic. I told them to appeal in accordance with Civil Procedure Act 2007 and other laws.

I left the court in an ever chaotic environment, as one of the former local wrestlers threw up a solid built 2nd Lt. Police officer and nearly strangled him. It was no long law reasoning working but the urge of physical confrontation. He nearly crushed him as he fell harder at the back. I saw cap flying in the air to three metres away. There was another middle age man screaming and running around madly looking for a gun to settle this matter with bullets.

He was crazy for it was said later that he sold three cows in the village of Cueibet to make business in town. So everyone was escaping from the court premises and the alleged judge had to lock himself inside the room. The room door was metallic door. It was like a movie.

It was like the judge had gone for self-imprisonment. Then he peeped through a shattered dirty door. One of the accused who had approached me before I crossed the road confidently said they had created a comedy which was to indirectly force the judge to jail himself. Immediately I left the court and went straight to the office. I shared the story with Geu Kuol and Counsel Anuol. The following day, I found the story in the newspaper The Juba Monitor. The cartoonist seemed like he was watching the court drama. He depicted a judge who asked the prosecutor to hurry up with a trail of dollars falling from the box.

The caption read 'There are no witnesses, please hurry up'. I took the paper to our office which was located along the Tumbura road. Three months after I lost the case in county court I met Bul. The inquiry put before him on the same matter got a reply that the matter in the court of appeal had the possibility of being lost too.

The case dragged on for one year, until I heard from the two defendants that one of the powerful ministers had intervened or interfered in their favour. I was going to the same court when I bumped onto them. In October I had another client whose brother came to me on his behalf since he was already remanded in custody. The defendant was accused by the landlord as alleged motorcycle thief. The motorcycle was under the care of someone on rental. The bike was wrongly parked at the corner of bamboo fenced compound. Then someone whose description fitted the accused escaped with it around six thirty pm. The incident happened in the Juba suburb Newsite.

The next day, the defendant aged twenty one was picked up by owner of the motorcycle, a soldier. The owner said they had to barricade the road and took in two people on the descriptions that they later released. Then the one inside the prison fitted the description.

Tall, giant and big red eyed were the descriptions given by a young lady who had been taking shisha tainted with cannabis. She admitted to her smoking spree and said that three of her friends were nonetheless sobered. She was the key witness in the prosecutor's case. Her evidence could have tipped the balance against my client had it not been because of the credibility or probity issue hanging around her neck. She was called to give evidence. She was a stereotypical depiction of the emerging urban dwellers. Her skin had spots which looked like lesions due to bleaching. The judge was waiting for the prosecutor to open his case. The prosecutor's argument came first before I was allowed to speak and after that both cross examinations were completed at the same time.

While she confirmed her description of the physical appearance of the accused, she failed terribly to answer specific questions. She did not know the colour of the motorcycle during examinations or if she was not under shisha at the time. And how well could she identify him among class of big eyed Jubians?

Her failure on description of the colour of the bike had made her narrative inadmissible, before the eye of law. She could not have precisely provided a detailed description of the accused since the act was not committed at night or done under threat.

To make matter worse, she might have been under the influence of illicit drinks. The same judge ordered that unless the accused paid the equivalence in kind; the accused should be remanded in prison forever. His clerk who sat next to him was acting in a funny way. He was not respecting the court etiquette. He interrupted the judge several times. It was another miscarriage of justice. Feeling defeated, I advised my client's brother to appeal the judge's decision. His brother acting under immerse pressure was willing to pay SSP 5000 to the court.

But I was uncomfortable with the clerk sitting next to the judge. He was a short and solid guy in white suit and yellow trousers. He was whispering to the judge and behaved abnormally. He was uneasy with himself. He might have had the possibility that my client had no case to answer since the prosecutor's witnesses including the owner of the motorcycle were not proving anything. The owner was not helping his case either because he could not provide genuine ownership documentation despite the judge referring to everything as exhibit number 01. Nevertheless, in my legal practice in Juba, I found good judges such as Francis, Lucia, Thok and Gerri.

Judge Thok was in second grade. After few years in the judiciary he decided to submit his resignation letter to the President who had appointed him.

They had to do the resignation letter submission together with another Judge Philip. It was on 5 December 2015 when I encountered Judge Philip on a certain case involving an unlawful arrest of a British citizen over a manpower service contract he signed with a second party. The second party signed another contract with a manpower provider. According to my client, the provider did not pay the employees and had left the country. Th employees wanted him to pay or put arrest.

In the process of exchanging plaints between plaintiff lawyers and our legal team from Awatkeer Law Chambers, the Judge ordered an arrest warrant based on the request of the plaintiff lawyers. The application was heard and determined in our absence.

I was very busy that day and I had to send junior lawyer Aguer Chol to make inquiry only to be surprised with the plaints. He rang me and narrated everything and as a result I rushed to the court.

I made an application for our client to be released so that we could file his affidavit and for us to reply to the plaintiff plaint/claim and statement of claim. This argument was rejected by the judge and instead our client paid over forty thousand dollars to the court account. I objected to his reasoning and argued that since he has not determined the case at trial and hearing, the payment should be the final if at all he found my client lawfully responsible for breach of the contract. My persistene on due process caused me trouble beginning with the possibility of the Hon. Judge having been bowdlerised by foreign English accent.

In most of the courts, the two languages of English and Arabic are used interchangeably. The judiciary had just departed from the use of Islamic penal code Sharia law. This sometime causes more confusion on which one should prevail between Sudan's previous precedents and common law precedents.

In collaboration with some ill trained lawyers, the judge was convinced to arrest a car dealer unless forty-six thousand of dollars were paid into court account. It was about ten forty five when one of the trainees I sent came to the office and narrated that he was served with plaint (claim and statement of claim). I put on my suit and rushed to the judge with an application to allow the client out of prison so that he replies to the claims from the plaintiffs. The judge response was negative.

He threatened to arrest me and ordered my arrest. He took a blue pen and wrote in Arabic and drafted a charge of contempt of court under section 70 of Civil Procedures Act 2009. He was breaching the constitution for a fair hearing on any criminal case. Nevertheless, he knew what he was doing was unprocedural for any crime must be tried before any conviction is recorded. He felt too small and his inadequacy was taking a toll on him.

He ordered my arrest on the allegation that I was making nuisance in a judiciary. He took the legal argument for noise which had irritated him.

There was no any evidence to support his allegation other than a serious misunderstanding since the Judge had ordered an arrest against one of foreign contractors with him not being give the right to be heard or presumption of innocent until proven guilty.

Together with my client, we were locked inside for nearly two days before the Bar Association exposed the judge's lack of legal analysis and gross misconduct.

He was also violating Section 3 of the Advocates Act 2013 which bars the arrest of lawyers without authorisation from the Minister of Justice or the Chairman of the Bar Association. He asked that I should write an apology which I refused before the Chairman of the Bar Association requested it. There was no trial in both cases and Judge won for the money was paid to a court account so that any form of legal battle was permitted. Having been released after 48 hours, I withdrew from the case.

By April 2015 I got another specialised assignment with the oil industry. I informed Daniel Anuol about the latest development and I quit legal practice and moved to Paloch. This offer was in response to a one year old comprehensive interview I did way back in 2014. The position was for someone with a legal background or a lawyer. The first positions were given to two lawyers and my name was filed with the Ministry of Petroleum with other four applicants pending further engagement. Tethered down by the legal profession, I almost forgot about it until one of my friends alerted me that I should check with the company.

I went and met one of Human Resources Managers had to take into consideration my right to employment when records had it that I had passed the interview. This appointment came after I had concluded the Intergovernmental Authority on Drought (IGAD) independent research on South Sudan Security sector reform. The IGAD is a regional bloc which comprises of over seven countries in the Horn of Africa. The research which I conducted in collaboration with the Republic of South Sudan Ministry of Foreign Affairs and South Sudan Embassy in Ethiopia was on the Relevance of Security Sector Reform in South Sudan and other crosscutting issues affecting the security sector transformation.

This contract came without my knowledge. I was later informed by my friend Moses who was working in the Ministry of Defence and Veteran Affairs and Army Counter-terrorism unit that he had submitted my academic credentials to IGAD. Then I and Chol were called to the South Sudan Foreign Affairs office which sits in one of the old complexes built during the reign of Gen. Joseph Lagu and Justice Abel Alier Kuai Kut. The director dealing with International Cooperation suggested that we should be flown to Addis Abbas to attend the interview, a suggestion which I later objected to as a waste of money.

In between 2005 to South Sudan Independence, the number of foreign trips taken by government officials were so many that if South Sudan developments were to be based on knowledge and exposure those officials made it to developed countries, then the country ranking below the poverty line would have been uplifted with Agricultural theories and practices learned in China, Europe, America and enormous African emerging economies.

However, the expeditions for those whose exposure was new had unbelievable aftermaths. For instance, a friend of mine who arrived in Canada in 2000 was reported to have refused to use the toilet inside his room because he perceived the bowl as a silver plate. He had to take another orientation before he joked that he and his friends had started using toilet for the first time. He was had lived in the refugee camps most of his life. They had got used to latrines and forests. The toilet had it own management and usage he would say.

Chol, Kulang and I left without agreeable conclusion. After two days, I was called back to the Foreign Affairs office for an IGAD independent consultant contract. The terms were that I must meet social expectations and the standard of academic research. I was allowed by the contract to appoint assistant consultants. Acting under that provision, I was privileged to appoint Anthony Ongem and Legal Officer in the Ministry of Justice Deng Mayom.

Deng was educated in South Africa while Ongem was UPFD officer with degree from Gulu University and had vast experiences in Somalia working under Africa Mission in Somalia (AMISOM). After two months of research, I submitted the document to an IGAD secretariat in Addis Abbas.

I got a notification that the team would review the document and a delegate would be sent to Juba for a validation workshop. I was asked to invite members of the organized forces such as the Army, Police, Prisons, Wildlife, Judiciary, Foreign Affairs.

The IGAD paid for the workshop and by eight thirty am the delegates arrived at the Shalome Hotel near Juba Airport road. As matter of procedure, the meeting was opened by a South Sudan senior government official. The secretariats of IGAD made some remarks before my team was asked to do the presentation. The delegates deliberated over the document which incorporates South Sudan post-war dynamic security environment and non-traditional threats of intercommunal violence, poverty, rebellions, natural disasters and political squabbling within the ruling party. The paper documented the messy divorce from Sudan and pugnacious along the borders with Sudan.

The bloody conflict over land seized by the Sudan's government had been rocking the south-north politics. Everyone was praising us as South

Sudan young researchers with government officials promising as usual to support research institutions in South Sudan. But not all of them were happy and one of the senior judges became upset. He accused our team for question the judiciary and particularly the Chief Justice of the Republic. He was the justice of the Court of Appeal, the second court above the High Court and below Supreme Court. The issue I highlighted was with respect to the role of other courts in the transitional constitution. I had the opinion that the provision is too vague for the people of South Sudan have the rights to know that they could go to the courts whenever one had a case. The language of the constitution should be very clear.

Like children, the judge called the two of us aside and angrily yelled at us: "Why do you young people like to question the integrity of the government institution?" Since South Sudan is a country founded on collective efforts and no one man's enterprise, I argued that there should be no consolidation of power over what is considered as an individual's absolute inherent rights. I told the judge that what he alleged had no substance. We were now talking as lawyers for judges are lawyers picked among the best to administer the rule of law. Everyone except him in the workshop even the most feared officers of the security services had no issue with the document. They even contributed effectively by providing crucial details left out regarding their mandate and operations He alone was trying to defend what was missed in the people's constitution.

He relaxed his emotions and decided to assist us incorporating into a document what those other courts were. After five minutes as Deng was doing the writing, he said the other courts are special courts. These courts are within the power of the Chief Justice who has right to constitute such courts.

We did as he instructed, but he had to question how and why the IGAD had recruited us in his closing remarks. Everyone was speechless.

CHAPTER 21

Working in Melut County

WHILE IN PALOCH, I was in charge of logistics with John Makoi, Mathew Ayok, and Michael Miyom. Paloch is suited in Melut County which is bordered by the counties of Manyo, Mabaan Baliet, and Renk. It hosts foreign and national workers. It is also surrounded by small human settlement. It is one of lowlands in Upper Nile region. Sometimes it has savannah grass sprouting up together with short acacia trees.

My being in Paloch would not have been possible if not for Procurement Manager Ayong Bol. Before I moved to Paloch I met the good people of Ajang Duot Bior, Thon Deng Ayong and Martha Ajak Majak and Abiong Kuol Deng. While there, Thon would make a call to check how I worked with the airport's busy traffic. It was the rainy season especially having been posted there in midst of month of May. Thon as a lawyer had graduated in the same year with me in 2010.

Thon was a workmate and a native of Akoka which is at the extreme south of Melut County and would sometimes urge me visit his village of Akoka. He said that I might enjoy fishing voyages on the White Nile bank. Thon said he was born in Koka and had witnessed the brutality of the civil war. He said one day his wooden boat was missed by missiles fired by Sudan Army Forces who were based at Kodok.

Akoka County is located north east of Malakal town. The people of Akoka are called Dongjol which is one of the Padang Dinka branches known as the Dongjol Ayong. It's name was from Ayongdit, who is also an ancestor of Ayueldit. Akoka people had a successful record of resistance against the Turko-Egyptian and Mahdi jihadist dominance. They used to call Arabs jur ayul because its local warriors in one of the battles killed an Arab man with too much hair covering all his body.

During the Anglo-Egyptian rule for instance, they were administered under indirect rule. This was made possible because the people there were well organized through the traditional spiritual order. Another story Thon cited was that, the churches' missionaries who came to the area began to establish the church at Rom northeast of Akoka. The church caused a dispute with a local spiritual leader called Ajang Chol (Ajang Tuobil) and had to be abandoned in favour of the traditional leader whose influence continued to reign.

During the pandemonium of the post-independence era, the Akoka County constituted an integral part of Sobat Rural District which comprised of (Akoka, Baliet and Malakal). Dongjol Dinka participated in the independence of South Sudan from Anyanya One to SPLA/M.

Akoka centre was captured by Anyanya One in 1970 led by Akech Yor and Ajak Guot (Ajak Nguat). The second attack was in October 1980 by Majok Yac a commander of Anyanya two forces. Majok's foot was seriously injured. He was acting under the command of William Geng Beek who later joined the SPLA/M in 1983. Because of its strategic location, Akoka County was one of the earlier liberated areas during the armed struggle although affected by SPLA/M 1991 split.

Its Anya-Nya One commanders who were later integrated into Sudan Army Forces with Addis Abbas Agreement included Commander William Deng Atem; Ayuel Mager; Bol Ayong Luong among others. These officers later became part of 104 and 105. Some of them became commanders of Mozlum Battalion meaning the oppressed and included Lt. Gen. Daniel Deng Alony, Sabri Choli, Ayuel Diing and Awan Tor among others. With Akoka between oil rich Melut County, Malakal and Baliet, the area experienced constant attacks from the Sudan Army Forces and its allies with an intention to uproot the civil population who were considered purely SPLA/M supporters. People like Brig. Gen. Thon Mum Kejok from Akoka were among the SPLA/M's best cadets. He was awarded an Diploma of political orientation which enabled him to serve his people as a Minister in the Upper Nile government. Other notable sons of the area included the former commander of northern Upper Nile Gen. Elia Biech Khor appointed as a MP of August House in 2005.

Thon Deng Ayong as my best friend will continue to be a bridge between Akoka and people of Twiland. Historically our people have many commonalities from Ayuel and Kur naming as in Ajoung, Nyarweng and Ayual community. He believes that we are descendants of the great grandfathers of Ayuel and Ayong. While I was living most of my life in the rebel held territories, refugee camp and in the west, Thon spent most of his life both in Akoka and oppressive Sudan of Khartoum. In Khartoum, he joined students' politics and was among the students who used to engage the police in street stone fights. One day he said he was drugged by the police with a lachrymator agent. He would narrate how the brutal police forced him inside a vehicle after he was nearly run over by the same police car. In those days, if one dies while in Sudan police hands, there was no accountability file. The file was closed upon someone's burial. He said his friend Chol Thon Kaak was murdered by the police and no one was held accountable.

He said in Khartoum, they had no option of joining the liberation as they were confined and not allowed to leave and hence sometime vented their frustration against the police. He continued that anyone who had read Sun Tzu's theory of victory would agree with him that victory has many forms other than military invasion.

There was concentration around the oilfields at the time I arrived. The Melut County's headquarter was destroyed by the forces of Johnson Olony and Unity Oilfields was among the strategic fields destroyed early on by the rebels of Riek Machar and they were eyeing Melut County. It was on 18 May, 2015 that I arrived in Paloch. It is a small lowland several miles from Renk town and on the east bank of the White Nile. At that time, there were massive movements of troops in and out of Paloch as two fighting forces tightened their grips over the oilfields. My job there was to coordinate all logistical aspects of oil production. These would include casing materials, food items, and production materials and lube oil.

I would make sure I was already at Paloch airport as soon as cargo left Juba International Airport. I would inspect the cargo and ensure everything was correct. I was trying something different from my work in Bilpam and legal practice. I was doing what I had never imagined but it was funny. I would interact with crew members and soldiers manning the airport. I would ask the crew to provide me with the list of items. Then the local loaders which were underpaid would be called to remove all the items from the plane before logistic vehicles dispatched them to various stores at Paloch or Adar.

I would diligently check that all the items had been delivered before I signed the delivery note. Then the crew the majority of whom were foreigners would go back to Juba provided they arrived there before six pm. Only aircraft with technical and mechanical problems stayed at Paloch.

Being there during the time of war reminded me of bad experiences of the South-North civil war. While at Paloch, I witnessed suffering and exposure to shocking experiences as the reasonable number of soldiers wounded in various battlefields around Kaka, Malakal, and Thiangriar, Kaka, Wadokana and Baliet began to increase. The airport was a transit station for the wounded waiting to be taken to Juba.

With no proper attention given to them or any contingency plan from their bosses at Headquarters, the junior officers in charge of their wounded colleagues would have no option, but to negotiate with cargo crew for them to accept their comrades rushed to hospital in Juba. From Juba to Paloch was about two hours of flight with cargo. Some of those planes were designed for carrying consignment and had little oxygen. In the downstream and upstream of oil production, there are chemicals used to stimulate the crudes and this means that the very cargo planes used to carry people were sometimes used to carry chemicals intended for production.

It was those cargo planes, which many people escaping the conflict in the Upper Nile were using. Those who would survive the pains would make it to Juba with cargo. Common too were families separated if the pilots decided to close the backdoor as people struggled to force themselves in. The airport security had to take care of children stranded at the airport as the plane took off with the mother leaving behind vulnerable children. Some of those cargo planes were rusty on the inside and twice as old as I was.

Achuil had to survive when their plane tyre burst in midair before hitting the ground harder causing a deep thin line on the runway with bulged metallic objects. I rushed to them and all were safe. I have known Achuil and another person who preferred to be called John since the day I arrived in Melut. They were South Sudanese working with a foreign crew. I boarded the plane and while there, I asked Achuil whether it would be a good suggestion for him to look for job somewhere for his safety was more important than working with a plane with a rotten and nearly disintegrating body. Then he smiled, before he confidently said they had had to fly to Juba sometime in March with only one wing working and they had made it safely.

I took pictures of the aircraft and shared it with aviation security and other security personnel. No one was willing to pay attention since the planes were alleged to have stemmed from Zol Kabier. Since the day I stepped into Juba from exile, I used to hear the word Zol Kubier without understanding its impact until a friend of mine who is Arabic conversant translated it as an equivalence of the biggest. Then I had doubt whether we human beings are really the biggest.

In our human kingdoms, Jubians would reason that, there are big fish different from the prophesied heaven kingdom where spirits from either the rich or the poor are said to be equal. I insisted on asking why such planes were allowed to fly in the airspace of South Sudan. To the contrary, no one was willing to talk to me even some of the soldiers manning the Airport. They were not willing to say anything about the plane for fear of reprisals. No one had energy to discuss the issue of safety.

Still pondering looking for answers, that day ended with no tangible solution on what to be done to aircraft that were not airworthy. It was time to go and I had to wish my friend Achuil a safe flight back to Juba. I handed him a signed original copy of the documents as he took another Soviet Union made aircraft from World War II. Then the plane took off and went straight into the dark cloudy June weather of the Upper Nile.

In the evening I would recount the pain of daily experiences. There were days I had to reconcile the past with the future South Sudan. Outside the operational base camp in Paloch were small hut made of grass and dirty barrow pits full of water. The young ones would swim inside and cases of children and drunkards drowning there were many.

Paloch was my life's missing link. It was an amazing experience exploring the oil facilities, getting connected to ordinary people and going around evaporation ponds where chemical water was sent back from Sudan to South Sudan after the oil was filtered. Paloch is typical of the bush life I saw in the 1990s. In the evening I would take a fifteen minute walk. On both sides of the road to Adar are located small villages. The people who lived there depended on water tankers. There were no taps and water was left by the road side in containers that birds can drink out of.

In the morning I would go back to the Airport. And when I found starving soldiers, I would immediately order my driver to bring food for them from the operational base camp (OBC).

The logistics lay with bosses in the Army Headquarters and sometimes they had to wait for supplies from Juba. This issue of inequality like us eating delicious food made of fish and chicken while the soldiers who had just arrived from Juba heading to the frontlines ate nothing nearly caused shooting one night from a drunk soldier.

It all started at five pm. It was when the security at the gate began to restrict intruders from accessing the dining hall. The entry was by work identification cards. The soldier was not allowed to get in. He had no work identification card. He only had the right of way to the compound but not to the dinner table. He argued against everyone and bitterly objected to such restrictions. He was told the orders came from above and that they were sorry.

His reply was that he would come with his legitimate national identification card (ID) anytime in the evening. He took off. He was seen going to the Paloch market where he charged himself fully with local made illicit alcohol. We were watching the news on the South Sudan Broadcasting Corporation (SSBC) with the likes of Lueth Gak Biar, Arok Aquin and other two old men on the division of South Sudan into small units of thirty two states. Then we heard the sounds of gun shots.

The man shouted: "My identification. This is my identification card," welding his AK47. After few minutes, he had the possession of another machine gun at a small outpost designed for security protecting the oil workers. Then he began to shoot into the air. Everyone was almost running away from the room before I calmed them down. I told them one must read the direction of fire and its intensity before deciding on an escape route. They listened and sat down.

About two seconds, the gunman saw three guys from among the security guards who came out from their room adjacent to my room and directed his fire against them. He released about three rounds. The first two rounds hit the power station and one nearly missed an old man called Koch. Shaken by fear, Koch was standing confused in the middle of the room at Block G.

Then I ordered that he should sit down for a while but after about few seconds, the bullet hit my room up above the door.

After ten minutes, the renegade escaped southwest of our block. He was an active shooter in the night. The commander in charge of the forces organised around the base were ordered to apprehend their colleague. After ten minutes, he was subdued without a shoot out or resistance.

I knew that one who engaged in such an act of indiscipline did not know that their conditions are concerning to some powerless quarters and that he was expressing his anger against the wrong people.

At the airport, I was also trying to interact with some of them and those we knew each other back in the days. People like my cousin Mangok Garang Juach were in division two. He was based around Juba before he was flown to Paloch with his colleagues to defend Malakal. While at the airport, I recognised him. I went straight to him. He told me they had a few hours and were waiting for another cargo to take them to Malakal. I rushed to my room and brought in some fresh milk including yogurt for him. I provided him with sugar. This was the third time I had met him since he left our village in 1987. He has been in combat all his life and being among the experienced Division 2 soldiers and older than the 2013 crisis induced recruits, they had instructions to defend Malakal and its adjacent localities.

As he said, they left very early in the morning. Some of his comrades whom I met never came back only to be remembered for their legacies. Mangok was there dug in around Malakal without time to see his family until he was granted permission when he received the message of the passing on of his brother Deng.

My seven months in Paloch gave me profound experiences and great disappointment as well. For instance, I able to understand fully the dire conditions of the people living around the oilfields. A majority of them live below the poverty line despite sleeping on the top of the natural resources bequeathed to them by God and their forefathers of Ayongdit and Ayueldit.

The people there are noble in all aspects of their social life. They are honest people capable of absorbing all challenges. When I encountered young women carrying grass or firewood on heads by the road side, I saw Aliet women in them.

Then I would order my driver Nyok to pick them up because the cars we were comfortably driving came from oil money. It was very rare to complete the journey to Adar without finding people barefoot on the road to either Paloch Market or a small settlement around the base of oil facilities.

The numbers of barrow pits were more than the schools and hospitals and level of local employment. At the front of our gate was a small camp named zero. It was a symbol of the living conditions of the people in the oilfield. Though zero is a number, below zero is negative. That was exactly what the Sudan government had done in Melut County.

Inside our camp was a giant mosque. It was a definition of Paloch and a symbol of a religion which was used as a tool of oppression. I rarely

ABOVE: Figure 26 view of Oil Facilities inside the car with Driver Arok Aquin 23 May 2015

BELOW: Figure 27 first time in Khartoum to attend buffer zone Pipeline repairs Negotiation with BAPCO Sudan, with Team Madut Akech, Dr Samuel, Corrosive Eng Padiet Deng, Sam Upstream and downstream coordination and I

went closer to it for any government which feeds ordinary people with religion or empty promises is devoid of logic and does not deserve to exist. Indeed, the Sudanese government was no more. The people there had fought for their rights and oil production was run by their government, sons, daughters and foreign investors. If there was anything which was not fixed, Sudan would argue that they had left the mess to the owners of the land. On 20 December 2015, I was recalled to Headquarters and my job title was changed to specialist in minor reshuffling. Seven months later in mid 2016, there was fighting in J1. This sadly claimed the life of my friend Jacob Diu Jok.

It was ten years before in Queensland when I had met the late Jacob Diu. He was the SPLM Chapter chairman in Queensland. He was a committed nationalist, whose love of his country brought him to Juba. He was a father of beautiful children who continue to reside in Brisbane. I used to attend SPLM activities organised by him.

It was around 2011 that he left Australia for South Sudan. I met him the day he was killed. Akoy who was President Office Manager and I persuaded him to leave the office as everyone was told of the deteriorating and unpredictable city security situation. His reply was that he had lost his nephew and would take the company's ambulance to facilitate the burial of his relative. We left and I headed to my rented apartment in Gudele One. The fighting started in J1 as the president and his deputy were adjudicating over national issues.

Jacob, his son and his driver were killed. His son was studying Engineering in India and was back for a few weeks in Juba on school break. By nine pm, there was a search by the security manager and an attempt to call his number failed. His body was found in the mortuary. His death was painful and unbearable and having met Jacob and interacted with him on many occasions, I was devastated. It had been two days before when I had tea with him at the Quality Hotel. The next day, everything was like Somaliland street fights and I ordered my family to leave Gudele One with immediate effect as the rebel fighters had flanked from the south of our house and were closing in from the southeast.

As Garang Alaak, my wife and I and some of my relatives left the house, the fighting ensued near Lou clinic. We woke for thirty minutes as I was carrying my little daughter before we made it to Thiongpiny. We proceeded to the house of Maj Gen Kulang Mayen Kulang and were warmly received by his wife. We slept there for one day and we decided to move to Suka Theta. Three days after the fight, I went back to assess the house left in the aftermath of Juba's second disruption. Juba was getting back to normalcy and I ordered that we move back to our house in Gudele the next day.

South Sudan was about to celebrate its fifth independence day. Like the Christmas of 2013, this celebration was again done with barrel of guns. The likes of my nephew Majok Athiei Abut who came from the United States for a family visit could not control his frustration. He was five years old. He was put up by his maternal uncle Akoy Machok Biar at the Paranoma Hotel.

Disturbed by sounds of guns, he grabbed his mother's phone and made a quick phone call to my number at ten pm. His request was: "Uncle, please order soldiers to stop shooting." I nearly cried for I was not sure what to tell him. Then I decided to tell him the impossible words of comfort that the soldiers would stop shooting as soon as their bosses had made the order. He hung up his phone.

His mother said he regretted coming to South Sudan though she was advised by friends it was not Uhuru on the part of Africa. The security situation returned to normalcy and when we escorted her to the Airport, she promised to come back. She said home is home regardless of the associated constant manmade headaches. My reply to her was that if you refer to headaches, we have panadol for all these and you should think of African pride all the time. She took off with Kenya Airways. Her six month old Achok was not scared and she was constantly smiling as gunshot sounds went crazy.

CHAPTER 22

Telephone Conversation and Face Meeting with Gen Kuol Manyang

Early on in 2013, I had the privilege of meeting Lt. Gen. Kuol Manyang Juuk. He was the Minister of Defence and Veteran Affairs. In the meeting, I asked for employment and his reply was that he wanted to professionalise the Army to improve the deplorable conditions of soldiers and their families.

That was the first time I met him. Then I met him again on 17 December 2018 with respect to East Africa Community Defence Cooperation Affairs matter. This meeting was organised through his executive director Maj Gen Adoor Deng Adoor. Adoor was the South Sudan People's Defence Forces (Defence Attache) to Washington, DC. The meeting came in the light of my appointment dated 3 December 2018, as South Sudan-Defence Liaison Officer (DLO) to Regional Defence Counter-Terrorism Centre of East Africa Community.

The Minister as I knew him those days in the liberated areas had unquestionable loyalty to the system and it was necessary to share with him some of the challenges facing us in the foreign mission. I told him that without people like him, Dr. John Garang De Mabior would not have managed to overcome the bush struggle and success. His orders on the recruitment of fighters and adherence to ideal of liberation struggle touched everyone including me.

In 1997, before some of us were taken to train centre near Kapotea, Maketh Diing, who was in group 50B had to abandon his studies when the Lt. Gen Kuol's agents rocked the Kakuma soil with a massive campaign. It was a call for volunteer fighters. In all his speeches with those I had attended back in Narus, he would strongly stress the consistent armed struggle to free ourselves. He continues to be respected across South Sudan and in the marginalised areas in the north. When I met him with Brig Gen. Kuai Deng and Capt Achak, it was on the issue of Defence's outstanding payment of South Sudan's financial contribution to the regional bloc.

He was told I was appointed but that I could not take my position in the East Africa community because of money. Then we were wondering whether he could persuade the minister of Finance, Salvatory Garang De Mabior to do something by clearing South Sudan's arrears in the region bloc.

We also discussed our recommendation on lifting South Sudan's sanctions among others which were endorsed by the Council of Ministers of East Africa Community Partner States. We told him out of nine recommendations of regional concern, six belong to South Sudan and therefore we urged for political leaders to do the follow up. South Sudan's economy was shaken by conflict and other dynamic disturbing factors and the Ministry of Finance and Economic Planning had difficulty on how to prioritise which issue to resolve first. After fifteen minutes we left and we proceeded to the Ministry of Finance ourselves.

At the 1st undersecretary office was a policeman standing seriously by the gate. He was busy opening the door to flowing influxes of people with all types of financial requests. I was among those people looking for the clearance of outstanding membership fees for East Africa unit associated with the Defence issues. The police officer was panicking upon seeing a certain Army general. I saw him pushing the crowd for the general to go. The privileges associated with ranks had made some South Sudanese act outside any military norms.

In my research in 2014, there was a recommendation on how to address the issue which began under the United Nations' big tent policy. This policy was based on the accommodation of various militants during 2005. In my new assignment in East Africa Community, I was based in either Nairobi or Arusha.

As part of my job requirements, I began to attend strategic defence meetings of East Africa community partner states in most of the capitals of the six partner states except Burundi. My first meeting of the East Africa Community on defence cooperation started in April 2018 when Juba was the host for the meeting. We received delegates from the Ministry of Defence from six partner states.

Figure 28 view of Malindi coastline 5 April 2018 with friend

It was South Sudan's turn to show its commitment to regional peace and security by participating fully on the issues affecting the regional bloc. The meeting lasted for two days and the farewell party was organised near the White Nile. I was the Master of Ceremonies. The meeting was graced with an Acholi dance which forced the current Chief of Defence Forces Gen Juma Johnson Okot to dance and sending everyone onto the dance floor. The party ended with a gift of crude honey. Each member of the delegation was provided with honey. In August, I and former Governor of Twic State Maj Gen Bona Panek Biar were sent to Kenya, Mombasa to attend the third round of the East Africa Community defence meeting. As usual, we spent two days deliberating on matters of concern to the region, Africa and beyond from non-traditional threats to armed conflicts in Partner States. The quest for sustainable peace in South Sudan and Somalia dominated most of the deliberations. The next day, we were driven to Malindi historical sites about two hours from Mombasa.

While there, our delegate was taken around the old building used by the sultanate of Oman Empire in officiating over their subjects including the slaves in 1861 until the reign of Sultanate of Majid of Zanzibar was ended by the British administration at the time the slave trade was made

Figure 29 inside the Palace April 5, 2018, Malindi

illegal. The head of delegates and accompanying staff were not the first guests in Malindi. The town was an epicentre of trade and civilisation in the 5th - 10th century. For instance, in the history books, high school students in Kenya were taught about the long distance trade which was taking place between the Bantu speaking farmers with Somali, Egyptian, and Nubians, Arabs, Persians and Indian traders.

The buildings which were scattered over 500 metres were made of stones. Some were ruins since most of its parts had already disintegrated. We were told it is being managed by the government of Kenya to preserve the ancient history. In my history subject class in secondary school, I was taught the impacts of Atlantic slave trade on the African Continent. Inside the building had tombs and big trees. In about a fourteen metre wide building divided into sections, inside glass containers, lay the tools used by the sultanate later referred to as Kings by the European explorers for conquering and subduing the subjects.

Figure 31 weapons used for slave raids at Archaeology of Gede

There was also a section of isolated corridors which were used to confine the slaves. In the evening we were driven back to Mombasa. The next day we left for Juba. Then our team which was mainly comprised of three personnel were informed by the East Africa Community Secretariat through the Republic of South Sudan Defence Liaison Officer Brig Gen Matur Dharui that the next meeting would be held in the first quarter of 2018. It was scheduled at the far foot of Mountain Kilomanjori in the Town of Mwanza.

Figure 32 Peak of Kilomanjori view from the Plane window 2 Dec 2018

We had a stopover for ten minutes at the base of Mountain Kilimanjaro.

It is a small airport for tourists willing to hike to the peak of mountain. Mwanza is located at the southern shores of Lake Victoria.

It has beautiful islands. The Saanane National Park Island is dotted with fragments of tall rocks.

It has thin and tall rocks separated from the island. Others may measure upto twenty metres high and standing randomly in the middle of the lake. It is ten minutes by small boats and is home to animals such as zebra, antelopes and lions. The male lion made a terrifying roar upon seeing us approaching the cage. We could view giant snakes such as pythons and different species bathing on the top of the rocks. We toured it for an hour before the guard called us to go back to our hotel.

We found old rocks, which I later assumed might have been formed millions of years ago when the earth formed imaginable creatures such as Lake Victoria, mountains and nearby islands. To connect from Kenya to Mwanza, one would see from distance the greatness of Lake Victoria. It was an amazing experience.

to Australia's Law School

Figure 33 at the heart of Mwanza before 3 December 2018 EAC Meeting, Tanzania

Figure 34 at the shore before boat ferried us to Saanane National Park Island, 5 December 2018

CHAPTER 23

Rwanda Genocide Memorial

It was early 2019 when our team was invited by the East Africa Community Secretariat on Defence Cooperation to attend a meeting in Kigali. I and two others officers were nominated and we left for Kigali via Nairobi. On arrival, we were taken to the Kigali Genocide memorial built for the remembrance of the 1994 genocide which nearly killed about two million people. It is the resting place for victims of genocide which was against the Tutsi.

By the time we arrived in Kigali, the nation had already healed and no one could tell who were Tutsi and Hutu. We found beautiful Rwandanese citizens who gave us the best reception from the Airport to our last day in Rwanda. We were honoured with salute and taken to the VIP section of the Airport. Our passports were cleared and we were driven off to one of the five star hotels. It was my third time to enjoy the privileges accorded to the diplomats. The protocol had changed so that I could easily get lost going to the ordinary section of the Airport before I was reminded to go to the VIP sectionn next time I came back to the country.

While I was doing research at the university on war crimes against women under international humanitarian law, Darfur and Rwanda were the focus of my interest. It was my first time in Rwanda. For anyone who had never been there would not believe how a country which was the laughing stock of the western media and used to shame Africans had transformed itself. The memorial is managed by the government to

Figure 36 East Africa Community Partner States meetings 2018 Juba & 2019 Kigali, Rwanda

remind humanity that if two brothers fight, folding hands for you to turn up on their burials is an equal crime of unconscionable measures.

Kigali as the capital of the Republic of Rwanda is surrounded by mountains and evergreen landscapes. These features define life in Rwanda. The weather is of great contrast with our hot weather in Juba. I nearly added on weight.

While on the tour of significant areas designated for visitors and guests, we were given instructions on how we should conduct ourselves. For instance, there was a signpost not to take photographs while inside the memorial building.

At the entrance was pictured the United Nations General who could not save the civilians running to their camps from violence. We were told by our tour guard that the United Nations Peacekeeping forces had to leave the Rwandanese to kill themselves. He said families were wiped out and ethnic cleansing nearly depopulated Rwanda. Because of violence, millions were displaced to refugee camps, he continued. The country became a butchery, one of the Rwandanese senior officers in our team added. The pictures attached to the wall in the remembrance of victims were shocking and devastating.

Inside there, we were lectured on what caused the genocide and showed the picture of the meeting which was attended by the two presidents of Rwanda and Burundi before their plane went down killing both of them. As a result of the assassination of the Rwandanese President, roadblocks were set up at night and a killing spree began. All the attempts by moderate Hutus in the government to calm down the situation cost them their lives instead.

We were showed portraits of victims which included female politicians, pastors and scholars who were murdered for opposing the genocide. Then we were walked through a section where ringleaders of different professions from pastors, politicians, students, witchcrafts, foreign agents to name them were exposed for having their hands in fuelling the genocide.

Then we were directed to a picture of an old woman. Her name is Zura Karuhimbi. This woman had the wits to protect the lives of genocide victims. She used her necromancy to save hundreds of people's lives. During the genocide, people were desperate to save their lives. Defenceless from the perpetrators, Zura, resorted to scare tactics to protect those who sought refuge in her compound from the murderers. Her depiction hangs proudly on the wall in the first quarter of the memorial building. We were told she protected the victims by scaring the perpetrators with magical power. She would say if anyone kills the victim the spirit of magic would send them to their own graves.

A few weeks before I arrived in Rwanda, the woman in question was reported by BBC to have died of natural cause. I thought it was fake news. In Rwanda, she was a heroine.

The Rwandanese were telling me that they had to move beyond the tribal line and forgive one another. They said that perpetrators and victims are living together side by side as citizens of the Republic of Rwanda. In Rwanda for instance, it is not permitted to call someone by tribe. Everyone is Rwandanese. The tour guard as matter of confidence showed me his national identification. He later added that they had abolished putting people's tribe, region and other details which revealed their personal identities. Then I told him South Sudan had an issue of tribal adherence like Rwanda and it was my hope that such brilliance needs to be exported to the rest of the partner states. I removed my IDs from my wallet. I showed its face to him. On its face is beautiful national flag. He grabbed it and gave it a close look before he said such précise personal details such as my place of birth contributed negatively to the genocide.

He handed it to me before saying that they had had enough from genocide and decent burials of the remains of victims would act as a dignified sending off the cruelty of mass murder. In his book, Dr James Waller's *Becoming Evil: how ordinary people commit Genocide and mass killing* was the Rwanda of 1994. However, we both agreed Africans should work together to move Africa beyond the bitterness which comes with violence extremism in wars. While on the plane back to Juba via Jomo Kenyatta International Aiport I had deep thoughts about Rwanda's orphans.

Then I was taken back to the Yugoslav wars. The International Criminal Tribunal for Former Yugoslavia (ICTY) proceeding I was watching with respect to the perpetrators of the war crimes and crime against humanity when I visited the court on 15 February 2009. Then I held my breath for seconds and remembered the book by Scott Peterson *Me against my Brother: At War in Somalia, Sudan and Rwanda*. This book along with *The Kennedy Curse: Why Tragedy Has Haunted America's First Family of 150 Years* by Edward Klein. It was twelve years ago in the summer of 2007 when I saw the two books aligned together in one row at James Cook University Library, in Townsville.

The curse of leadership and brothers slicing each other throats with machetes and swords as in Rwanda and Yugoslavia dominated my flight.

By two pm, the plane was already in Nairobi. Had the pilot not made an announcement, I would not have known passengers were leaving the aircraft. Then I proceeded to see my family. I found my children and mother waiting for me. I came to know her in 2013 before our traditional married in 2014. The next day I flew to Juba where I and Bishop Majok Dau and Bishop Henry Garang and Counsel Deng Akech involved in community intra dialogue for harmonious living.

www.ingramcontent.com/pod-product-compliance
Lightning Source LLC
Chambersburg PA
CBHW021403290426
44108CB00010B/371